Nano-Optics and Nanophotonics

The Springer Series in Nano-Optics and Nanophotonics provides an expanding selection of research monographs in the area of nano-optics and nanophotonics, science- and technology-based on optical interactions of matter in the nanoscale and related topics of contemporary interest. With this broad coverage of topics, the series is of use to all research scientists, engineers and graduate students who need up-to-date reference books. The editors encourage prospective authors to correspond with them in advance of submitting a manuscript. Submission of manuscripts should be made to the editor-in-chief, one of the editors or to Springer.

More information about this series at http://www.springer.com/series/8765

Takashi Yatsui
Editor

Progress in Nanophotonics 6

Springer

Editor
Takashi Yatsui
Toyohashi University of Technology
Toyohashi, Aichi, Japan

ISSN 2192-1970 ISSN 2192-1989 (electronic)
Nano-Optics and Nanophotonics
ISBN 978-3-030-71515-1 ISBN 978-3-030-71516-8 (eBook)
https://doi.org/10.1007/978-3-030-71516-8

This Springer imprint is published by the registered company Springer Nature Switzerland AG
The registered company address is: Gewerbestrasse 11, 6330 Cham, Switzerland

Preface to *Progress in Nanophotonics*

As the first example, recent advances in photonic systems demand drastic increases in the degree of integration of photonic devices for large-capacity, ultrahigh-speed signal transmission and information processing. Device size has to be scaled down to nanometric dimensions to meet this requirement, which will become even more strict in the future. As the second example, photonic fabrication systems demand drastic decreases in the size of the fabricated patterns for assembling ultra-large-scale integrated circuits. These requirements cannot be met even if the sizes of the materials are decreased by advanced methods based on nanotechnology. It is essential to decrease the size of the electromagnetic field used as a carrier for signal transmission, processing, and fabrication. Such a decrease in the size of the electromagnetic field beyond the diffraction limit of the propagating field can be realized in optical near fields. Nanophotonics, a novel optical technology that utilizes the optical near field, was proposed by M. Ohtsu (the editor of this monograph series) in 1993 in order to meet these requirements. However, it should be noted that the true nature of nanophotonics involves not only its ability to meet the above requirements, but also its ability to realize qualitative innovations in photonic devices, fabrication techniques, energy conversion, and information processing systems by utilizing novel functions and phenomena made possible by optical near-field interactions, which are otherwise impossible as long as conventional propagating light is used. Based on interdisciplinary studies on condensed matter physics, optical science, and quantum field theory, nano-materials and optical energy transfer in the nanometric regime have been extensively studied in the last two decades. Through these studies, novel theories on optical near fields have been developed, and a variety of novel phenomena have been found. The results of this basic research have been applied to develop nanometer-sized photonic devices, nanometer-resolution fabrication, highly efficient energy conversion, and novel information processing, resulting in qualitative innovations. Further advancement in these areas is expected to establish novel optical sciences in the nanometric space, which can be applied to further progress in nanophotonics in order to support the sustainable development of people's lives all over the world. This unique monograph series entitled Progress in Nanophotonics in the Springer Series in Nano-optics and Nanophotonics is being introduced to review the results of advanced studies in the field of nanophotonics and covers the most recent topics of theoretical and experimental interest in relevant fields, such as classical and quantum optical sciences, nanometer-sized condensed matter physics, devices, fabrication techniques, energy conversion, information processing, architectures, and algorithms. Each chapter is written by leading scientists in the relevant field. Thus, this monograph series will provide high-quality scientific and technical information to scientists, engineers, and students who are and will be engaged in nanophotonics research. As compared with the previous monograph series entitled Progress in Nano-Electro-Optics (edited by M. Ohtsu, published in the Springer Series in Optical Science), this monograph series deals not only with optical science on the nanometer scale, but also its applications to technology. I am grateful to Dr. C. Ascheron of Springer-Verlag for his guidance and suggestions throughout the preparation of this monograph series.

Tokyo, Japan
August 2010

Motoichi Ohtsu

Preface to Volume VI

This book focuses on various and mutually related topics in nanophotonics written by the world's leading experts. Following the elaboration of the idea of nanophotonics, much theoretical and experimental work has been carried out, and several novel photonic devices, high-resolution fabrication, highly efficient energy conversion, and novel information processing were developed in these years. Novel theoretical models describing the nanometric light–matter interaction were also developed. For reviewing and demonstrating this recent progress in nanophotonics, this book focuses on various and mutually related topics in nanophotonics written by the world's leading experts. The first chapter describes controlling the exciton dynamics using metal/dielectric interfaces for improving light emitting devices with quantum structures. The second chapter is devoted to describing a novel platform for light–molecule interactions by the combination of ultrashort-pulsed laser and plasmonics in the mid-infrared range. The third chapter describes recent progress in the development of plasmon-induced water splitting, especially photoelectrochemical water splitting. The fourth chapter describes theoretical and computational approaches to describe ultrafast and nonlinear optical responses in nano-materials based on ab initio time-dependent density functional theory for understanding phenomena in nanophotonics. The last chapter describes an overview of optical spectroscopy of carbon nanotubes with quite rich physics for future application in photonics. This volume is a valuable resource for readers and future specialists in nanophotonics.

Toyohashi, Japan
August 2020

Takashi Yatsui

Contents

Contributors

Satoshi Ashihara Institute of Industrial Science, The University of Tokyo, Meguro-ku, Tokyo, Japan

Yuichiro K. Kato Nanoscale Quantum Photonics Laboratory, RIKEN Cluster for Pioneering Research and Quantum Optoelectronics Research Team, RIKEN Center for Advanced Photonics, Saitama, Japan

Kaihui Liu State Key Laboratory for Mesoscopic Physics, Frontiers Science Center for Nano-optoelectronics, School of Physics, Peking University, Beijing, China

Shigeo Maruyama Department of Mechanical Engineeing, School of Engineering, The University of Tokyo, Tokyo, Japan

Hiroaki Misawa Research Institute for Electronic Science, Hokkaido University, Sapporo, Japan;
Centre for Emergent Functional Matter Science, National Yang Ming Chiao Tung University, Hsinchu, Taiwan

Ikki Morichika Institute of Industrial Science, The University of Tokyo, Meguro-ku, Tokyo, Japan

Koichi Okamoto Department of Physics and Electronics, Graduate School of Engineering, Osaka Prefecture University, Sakai, Osaka, Japan

Xu Shi Creative Research Institution, Hokkaido University, Sapporo, Japan

Takashi Takeuchi Center for Computational Sciences, University of Tsukuba, Tsukuba, Japan

Mitsuharu Uemoto Department of Electrical and Electronic Engineering, Kobe University, Kobe, Japan

Kazuhiro Yabana Center for Computational Sciences, University of Tsukuba, Tsukuba, Japan

Atsushi Yamada Center for Computational Sciences, University of Tsukuba, Tsukuba, Japan

Shunsuke Yamada Center for Computational Sciences, University of Tsukuba, Tsukuba, Japan

Chapter 1
Highly Enhanced Light Emissions from InGaN/GaN Based on Nanophotonics and Plasmonics

Koichi Okamoto

Abstract Optical properties of InGaN/GaN-based quantum wells (QWs) commonly used in light-emitting diodes (LEDs) are determined by the time–space dynamics of excitons generated by current injection and photoexcitation. Such exciton dynamics can be directly observed by temporal and spatial-resolved spectroscopic techniques based on nonlinear optics and near-field optics. Nanophotonics using nanostructures of materials and plasmonics using metal/dielectric interfaces enable us to control the exciton diffusion and recombination dynamics and its emission mechanism and bring more efficient light-emitting properties.

1.1 Background of the InGaN/GaN-Based Light-Emitting Devices

Solid-state light-emitting devices have been developed and expected as new-generation light sources because of their advantages such as small, light-weight, long lifetime, easy operation, and saving energy. Since 1993, InGaN/GaN quantum wells (QW)-based light-emitting diodes (LEDs) have been continuously improved and commercialized as light sources in the ultraviolet (UV) and visible spectral regions [1]. In 1996, white light LEDs, in which a blue LED is combined with yellow phosphors, have been developed [2] and offer a replacement for conventional incandescent and fluorescent light bulbs. Typical device structures of InGaN/GaN blue LED and white light LED are shown in Fig. 1.1. However, these devices have not fulfilled their original promise as solid-state replacements for light bulbs as their light-emission efficiencies have been limited. The most important requirement for competitive LEDs for solid-state lighting is the improvement of their quantum efficiencies of light emissions. The efficiency of white LEDs has only just reached 100 lm/W in practical use and is far from the theoretical limit of ∼300 lm/W, and there is still room for improvement.

K. Okamoto (✉)
Department of Physics and Electronics, Graduate School of Engineering, Osaka Prefecture University, Sakai, Osaka 599-8531, Japan
e-mail: okamot@pe.osakafu-u.ac.jp

T. Yatsui (ed.), *Progress in Nanophotonics 6*, Nano-Optics and Nanophotonics,
https://doi.org/10.1007/978-3-030-71516-8_1

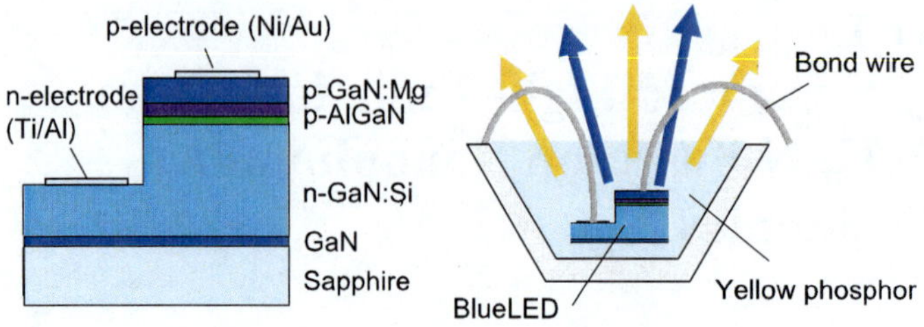

Fig. 1.1 Device structures of InGaN/GaN blue LED (left) and white light LED (right)

Figure 1.1 shows the emission wavelength dependence of the external quantum efficiency (EQE) of existing LEDs [3]. InGaN is a ternary mixed crystal and is able to cover the full visible spectral range by controlling the bandgap energies from InN (0.6 eV; 2000 nm) to GaN (3.4 eV; 365 nm). At the blue spectrum region, the EQE of the LED has achieved 75.5% at room temperature in spite of high threading dislocation [4]. However, such a high efficiency is achieved only in the blue spectral region, and it rapidly decreases outside this spectral range. For this reason, high-efficient InGaN LEDs with green, yellow, or red spectrum regions have not so far been achieved. High-efficiency red LEDs have already been achieved using AlInGaP-based QWs. However, it has been very difficult to develop high-efficiency green LEDs. This so-called green gap issue is also one of the serious problems preventing an alternative of LEDs to conventional light bulbs. Using AlGaN, which is also a ternary mixed crystal, the ultraviolet wavelength is available by controlling the bandgap energies from GaN (3.4 eV; 365 nm) to AlN (6.2 eV; 200 nm). In recent years, AlGaN/AlN-based QWs have been attracting much attention for their use in UV LEDs [5, 6]. The efficiencies dramatically decrease at wavelengths shorter than the bandgap of GaN (365 nm) and it is still difficult to develop high-efficiency LEDs even in the UV region. These issues of the green gap and the UV threshold are serious problems preventing the adoption of LEDs as alternatives to conventional light bulbs. The other issue is cost. The running cost of long-term use of LEDs is already better than that of fluorescence lamps; however, the initial cost is still expensive and this has delayed the spread of LED lightning.

The EQE (η_{ext}) of LED is given by a product of the internal quantum efficiency (IQE: η_{int}) and the light extraction efficiency (LEE: C_{ext}). The η_{int} values in turn are determined by the ratio of the radiative recombination rate (k_{rad}) and the nonradiative recombination rate (k_{non}) of excitons as shown in (1.1)

$$\eta_{ext} = C_{ext} \times \eta_{int} = C_{ext} \times \frac{k_{rad}}{k_{rad} + k_{non}} \tag{1.1}$$

Often, k_{non} is higher than k_{rad} at room temperature, resulting in modest η_{int}. Therefore, there are three methods to increase η_{ext}; increase C_{ext}, decrease k_{non}; or increase k_{rad}. Several reports have been published to improve from InGaN LEDs by using the patterned sapphire substrates and mesh electrodes [7] or making the photonic bandgap structures [8]. Much more effort has been placed into reducing k_{non} by growing higher quality crystals because usually η_{ext} of LEDs depend on the threading dislocation (TD) densities. For example, the TD densities were reduced to $10^6 cm^{-2}$ by using the epitaxial lateral over growth (ELO) technique [9]. However, η_{ext} of LEDs grown on ELO-GaN with longer wavelength emissions were not so improved, while those with UV-blue emission are well enhanced [10]. There have been very few studies focusing on increasing k_{rad}, though that could prove to be most effective for development of high η_{ext} light emitters. Controlling of the radiative recombination process of excitons is very difficult because it depends on many factors.

In this chapter, firstly, the relationship between exciton dynamics and emission properties is explored using various measurement techniques to elucidate the factors to determine k_{rad}. Secondly, methods to control the exciton dynamics based on nanophotonics and plasmonics to increase the k_{rad} and improve the emission efficiencies will be introduced.

1.2 Characterization of Exciton Dynamics and Optical Properties

1.2.1 Emission Rates and Internal Quantum Efficiencies

In order to control exciton recombination dynamics and improve the emission efficiencies, we have to elucidate and understand the important factors that determine the optical properties of InGaN QWs. Figure 1.3 shows the photoluminescence (PL) spectra and the Arrhenius plot of the integrated PL intensities obtained at various temperatures taken for InGaN/GaN LED wafers with blue, green, and amber light emissions [11]. These wafers were grown by a metal-organic chemical-vapor deposition (MOCVD) method on sapphire substrates. The active layer is a 3-nm-thick QW of undoped InGaN. To perform the PL measurements, a cw-He-Cd laser (325 nm) was used to excite the GaN cladding layers. PL was collected and focused into an optical fiber and subsequently detected with a multichannel spectrometer. The measurement was done at various temperatures by using a cryocooler with a helium compressor. The integrated PL intensities are constant in a low temperature region and then decrease gradually with increasing temperature. The exciton dynamics in the InGaN layer is well stabilized at a low temperature, and it is activated by thermal energy with increasing temperature, and easily captured into the nonradiative recombination centers (NRCs). In such a case, the η_{int} values at room temperature can be estimated by assuming that η_{int} is nearly equal to unity at low temperature (20K). The η_{int} values were estimated to be 35%, 30%, and 18% for blue, green, and amber LEDs, respectively. The η_{int} values including the values obtained in Fig. 1.3 were plotted against the emission wavelengths in Fig. 1.4. It was found that the η_{ext} value decreased with increasing emission wavelength, i.e., with decreasing indium (In) composition in the InGaN active layer. This tendency is similar to the emission

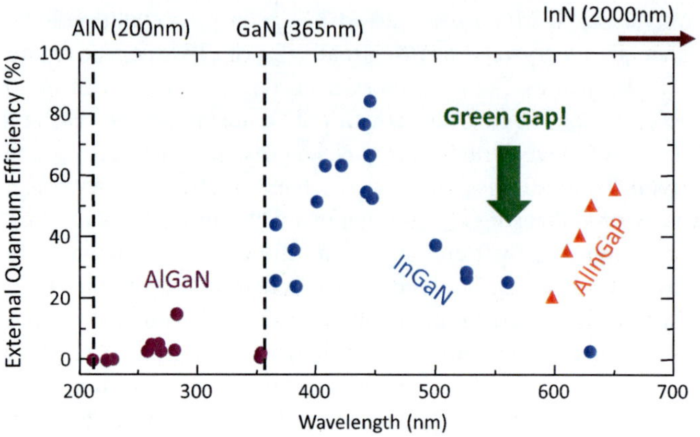

Fig. 1.2 External quantum efficiencies of various LEDs plotted against emission wavelength

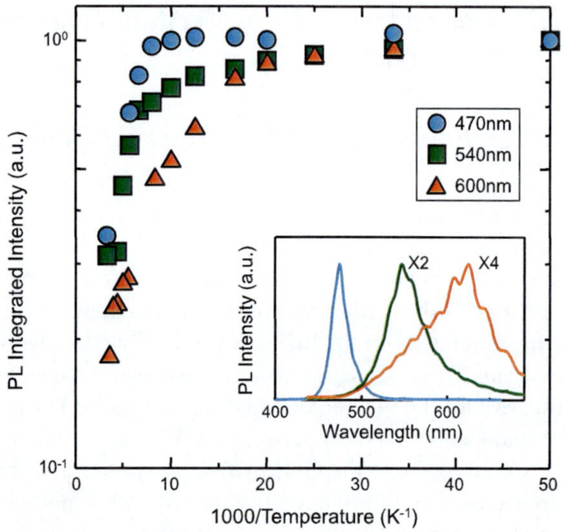

Fig. 1.3 PL spectra and temperature dependence of the PL intensity integrated over the emission spectrum taken for InGaN/GaN LED wafers with blue, green, and amber light emission. PL integrated intensities at 20 K were normalized to 1. Adapted from [11] Copyright© 2005, AIP Journals

wavelength dependence of the η_{ext} values shown in Fig. 1.2. In the UV region, η_{ext} become higher with increasing the In-composition. On the other hand, in the visible region (longer than 460 nm), η_{ext} decreased with increasing In-composition. This might be due to the increase in the NRCs, such as defects and translocations, due to the deterioration of the crystallinity of the ternary mixed crystal with increasing In-composition.

Next, we measured the exciton recombination processes by using the time-resolved PL measurements with picosecond pulsed laser, because the emission intensities are controlled by both the radiative and nonradiative recombination processes

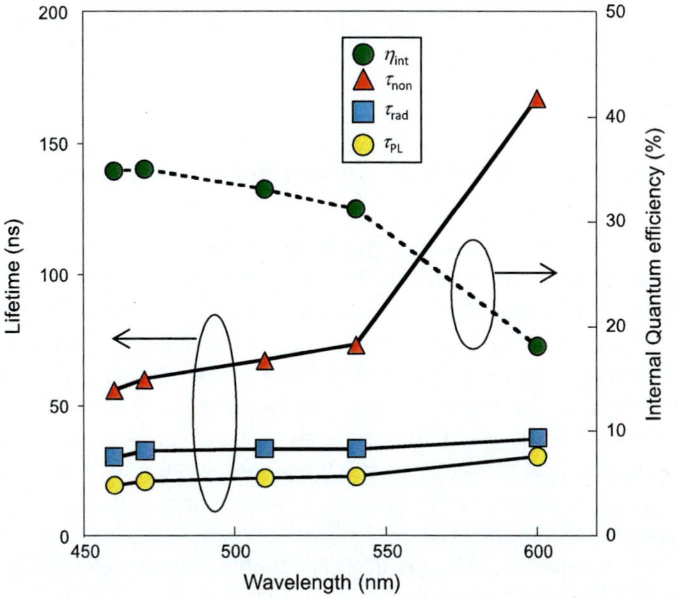

Fig. 1.4 Internal quantum efficiency and the PL lifetime, radiative and nonradiative recombination lifetime for InGaN LEDs with various emission wavelengths

as (1.1). The PL lifetime (τ_{PL}) obtained by the time-resolved PL measurements was also plotted in Fig. 1.4 [12]. It was found that the τ_{PL} becomes longer with increasing emission wavelength (In-composition). τ_{PL} can be separated to radiative lifetime ($\tau_{rad} = 1/k_{rad}$) and nonradiative lifetime ($\tau_{non} = 1/k_{non}$) by the following equations and also plotted in Fig. 1.4.

$$\tau_{rad} = \frac{1}{\eta_{int}}\tau_{PL}, \quad \tau_{nan} = \frac{1}{1-\eta_{int}}\tau_{PL} \tag{1.2}$$

It was found that τ_{PL} nearly equals to τ_{non}, and τ_{rad} of amber LED emitting at 600 nm is much longer than another one. As depicted in this figure, τ_{non} values were not so changed at 10 ns with increasing emission wavelength. These results suggest that the probability of nonradiative recombination pathway should be not so changed with wavelength. The probability of nonradiative processes is contributed by three terms, namely, the thermal velocity of excitons, density of nonradiative recombination centers, and capture cross section to the nonradiative centers. Therefore, the increase of τ_{non} with In-composition itself cannot solely be ascribed to the reduction of NRCs due to improvement of crystal qualities. On the other hand, τ_{rad} increase with In-composition. A dramatic increase of τ_{rad} was observed from 540 to 600 nm. This longer τ_{rad} causes the low emission efficiency at the longer wavelength region. The reason for this fact is difficult to explain only by the decrease in crystallinity due to the increase in In-composition, but the relationship with exciton dynamics needs to be elucidated in detail. This should be necessary to elucidate the factors to determine k_{rad} and to control them to obtain high emission efficiencies.

1.2.2 Optical Properties with Submicron to Nanometer Scale

It has become known that microscopic inhomogeneities of optical properties of InGaN active layers are very important for emission efficiencies [13, 14]. The microscopic spatial inhomogeneities of the optical properties should be generated by a fluctuation of In-composition by a phase separation in InGaN layers. The excitons in the emission layer of InGaN QWs are localized to the In-rich regions with relatively low energy levels, and the excitons are prevented from recombining through nonradiative processes [14, 15]. The relatively In-rich regions in the emission layer of QWs cause higher IQEs, and the In-rich regions are referred to as the "emission center." In-rich region in the emission layer can act as an emission spot for not only blue but also green-emitting QWs [16].

The other important factor is the quantum confined Stark effect (QCSE) caused by the piezoelectric field along the c-axis, which is the usual direction of growth [17]. Strain in crystal induces piezoelectric fields in the energy band diagrams of InGaN separating electron–hole pairs and decreasing the emission efficiencies. The QCSE becomes more effective with increasing In-composition in InGaN QWs, especially with green or red emissions and it causes lower IQEs. Avoiding QCSE is crucially important for the higher performance of LEDs. For example, Walterelt et al. fabricated the piezoelectric field free GaN/AlGaN QW grown on M-plane of GaN substrate and observed about 10 times faster spontaneous emission rate [18]. High-efficient light emission from InGaN/GaN QW grown on [11, 22] semipolar GaN substrate was also reported [19]. Local profiles of the QCSE due to the spatial inhomogeneity of the In-composition are also expected to be a very important factor of the optical properties. Therefore, characterizations of nanoscopic optical properties should be necessary to understand the exciton dynamics and improve the emission efficiency.

Cathodoluminescence (CL) [20, 21], confocal scanning laser microscope (CSLM) [11, 22], and scanning near-field optical microscopy (SNOM) [15, 23] are powerful methods to obtain optical images of InGaN/GaN in submicrometer scale. Figure 1.5 shows the typical images of the microscopic PL (a) and the CL (b) taken for InGaN/GaN QW with blue, green, and amber emissions at room temperature. Both images have microscopic inhomogeneities and the CL images have fine island-like structures with submicron scales due to the spatial fluctuation of the In-compositions in InGaN active layers [24]. The size of the island-like structures of blue emission is submicron scale, but it increases with increasing In-composition and become larger than 1 μm for the InGaN with amber emission. Such size should be much larger than the diffusion length of excitons.

Figure 1.5c shows the relationship between the PL peak intensities and wavelength of InGaN/GaN QW which have blue emission measured by the SNOM [15]

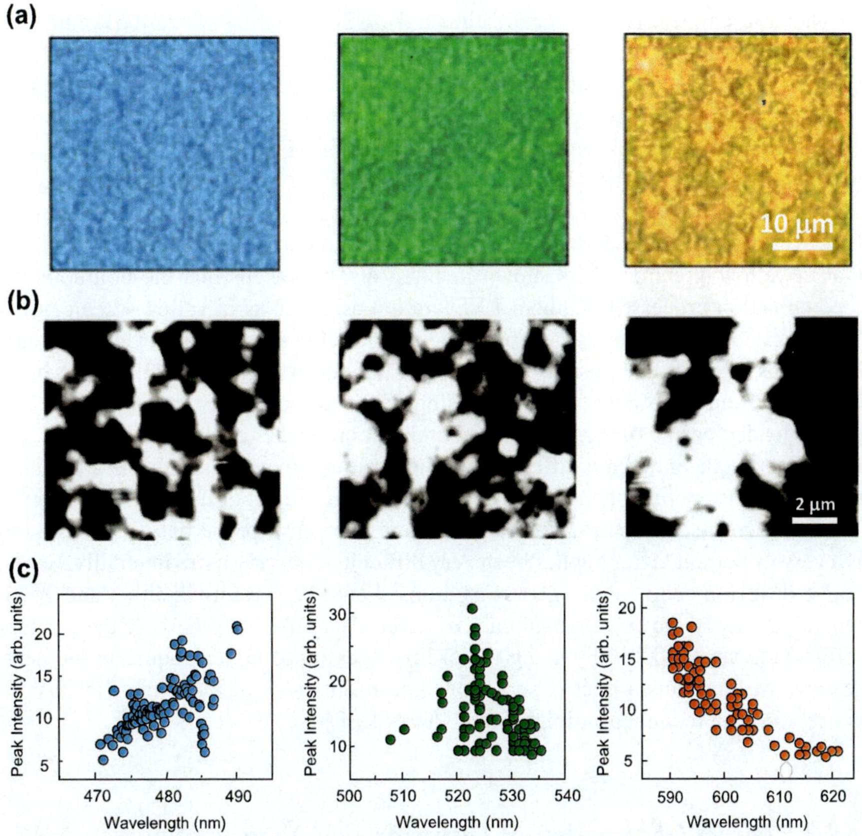

Fig. 1.5 Microscopic photoluminescence images (**a**) and Cathodeluminescence images (**b**) taken for InGaN/GaN QW which have blue, green, and amber emissions at room temperature. **c** Relationship between the PL peak intensities and wavelength of InGaN/GaN QW which have blue, green, and amber emissions. Adapted from [24] Copyright© 2009, IEEE Journals and [11] Copyright© 2005, AIP Journals

and green and amber emissions measured by the CSLM [11]. For InGaN/GaN with blue emission, a clear increase along the wavelength was obtained. This positive correlation has been well interpreted by using the exciton localization, which should be the main origin for high emission efficiency of blue emission. On the other hand, the correlation of InGaN with green emission looked normally distributed, while that of amber emission seemed to be decreasing along the wavelength. The negative correlation of amber emission is quite opposite to that of blue emission. This may suggest that the lower energy regions in the active layer of InGaN with amber emission should not be as effective as the localization center of excitons. We observed a similar tendency by using the CL measurements [24] and it should be related to the reduction of emission efficiencies of InGaN LEDs with high In-composition.

The different behaviors of localization and QCSE dynamics of excitons in InGaN active layers should be the reason for different optical properties with several In-compositions. The delocalization of excitons in larger In-rich areas and reducing the electron–hole wavefunction overlapping by the QCSE should be the main reason for the reduction of the emission efficiencies of LEDs with a longer emission wavelength. This reduction has been often explained by the low crystal quality of the high In-composition and activation of the nonradiative recombination centers. However, we have found that τ_{non} of excitons in amber LED is much longer than that in yellow–green LED as shown in Fig. 1.4. This means that the nonradiative recombination processes in amber LED are not as active as in yellow–green LED. We have also found that τ_{rad} in amber LED is much slower than that in yellow–green LED. This suggests that the reason for the low efficiencies of InGaN LEDs with high In-composition is not due to the increasing of the nonradiative recombination, but due to the decreasing of the radiative recombination by the remarkable QCSE in the InGaN active layer. Such a difference in the exciton dynamics should be attributed to the different spatial inhomogeneities and In-compositions of InGaN active layers. The submicron-scale spatial inhomogeneity of the optical properties observed here is a very important factor but has been very difficult to observe experimentally. Especially, diffusion coefficients (Ds) of excitons should be certainly a direct and good quantity to reveal the exciton dynamics in materials. However, to observe the exciton diffusion is very difficult by using the ordinal spectroscopic technique. In the next section, we introduce direct observations and discussions of the exciton diffusion process in blue to amber-emitting InGaN/GaN LEDs.

1.2.3 Direct Observation of Diffusion and Nonradiative Processes of Excitons

The internal quantum efficiency (η_{int}) of InGaN/GaN QW is controlled by both radiative and nonradiative recombination processes of excitons created by photoexcitation or electron injection. However, only a few investigations have so far been conducted to experimentally elucidate the nonradiative processes, primarily due to the difficult nature of such measurements. Transient grating (TG) method which is one of the third-order nonlinear optics is powerful tool to directly detect the nonradiative dynamics of excitons [12, 25, 26]. This method has been used to detect the optical properties of semiconductor at the very fast time region (nonlinear susceptibility [27], exciton dephasing time [28], time response of scattering [29], or quantum beat [30], so on) with the pico- or femto-second pulsed lasers. Haag and co-workers reported that the signal attributed to the thermal dynamics was observed after the signal attributed to the exciton dynamics in GaN [31]. In this section, we observe the spatial diffusion processes of excitons in InGaN/GaN-based LEDs emitting at the blue to amber spectral region by this method with a picosecond pulsed laser system.

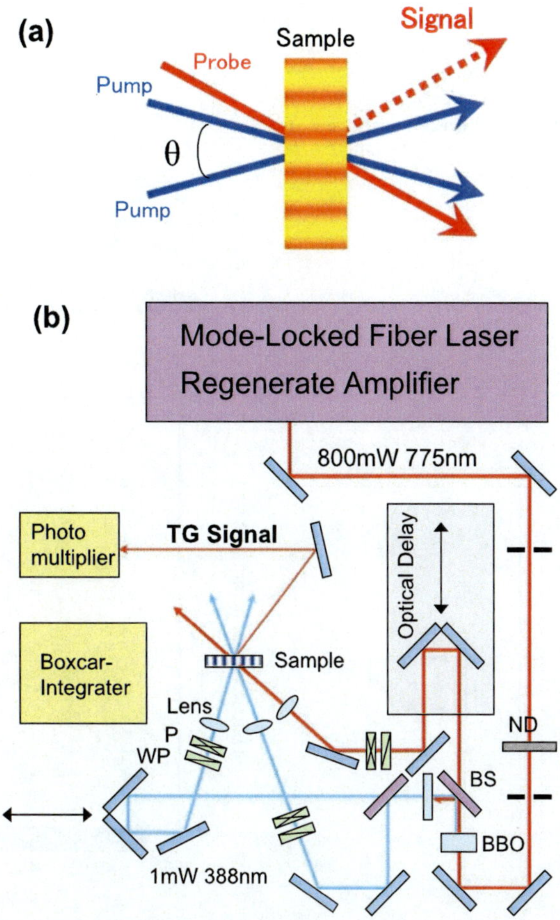

Fig. 1.6 **a** Configuration of the pump beam and probe beam of the transient grating (TG) method. **b** Experimental setup of the TG method with femtosecond pulsed laser system

The configuration of the pump beam and the probe beam (a) and the experimental setup of the TG method with picosecond pulsed laser (b) were shown in Fig. 1.6a and b, respectively [12]. A mode-locked fiber laser and regenerative amplifier system (Clark) were used. Frequency-doubled beam (λ_e= 388 nm) was created by the BBO crystal and used as the pump beam. Pulse width, power, and repetition rate of the pump beam were 500 fs, 1 mW, and 1 Hz, respectively. Pump beam was split into two coherence beams and crossed again at the sample with θ. Interference pattern was generated on the sample. The fringe space (Λ) was calculated as 0.7 μm by using the following equation.

$$\Lambda = \frac{\lambda_e}{2\sin\left(\frac{\theta}{2}\right)} \tag{1.3}$$

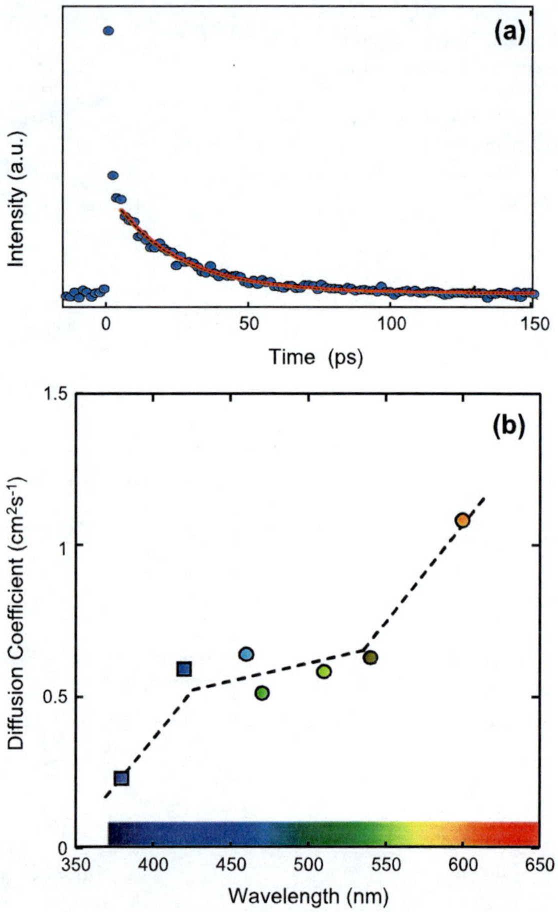

Fig. 1.7 **a** Typical time profile of the TG signal with femto-second pulsed laser taken for InGaN/GaN QW. **b** Exciton diffusion coefficient of InGaN/GaN QW with various emission wavelengths

The fundamental beam (λ_p=775 nm) was used as the probe beam with the optical delay unit. The TG signal was detected by a photomultiplier tube (Hamamatsu) and averaged with a boxcar-integrator.

Typical time profile of the TG signal with picosecond pulsed laser was shown in Fig. 1.7a [12, 26]. The TG signals decayed with a few tens picosecond. Similar signals could be observed by other LEDs. The rise and decay of this signal should be attributed to the generation and diffusion processes of the excitons. The rise component of this signal was similar to the excitation pulse shape. From the couple wave theory, the TG signal intensity is given by the sum of squares of the refractive index change (Δn) and the absorbance change (Δk) of the materials [32].

$$I_{TG} = \alpha \Delta n^2 + \beta \Delta k^2 \tag{1.4}$$

where α and β are the constant. At this time, Δk was negligible because the GaN has no absorption band at the probe wavelength (775 nm). Then, the intensity of the TG signal (I_{TG}) is proportional to the squares of the refractive index change (Δn^2). The time and spatial dependence of the refractive index change $[\Delta n(x,t)]$ are consisted by the time and spatial-depended exciton density change $\Delta N(x,t)$ and temperature change $\Delta T(x,t)$ as follows.

$$\delta n(x,t) = \left(\frac{dn}{dN}\right)\delta N(x,t) + \left(\frac{dn}{dT}\right)\delta T(x,t) \tag{1.5}$$

$\Delta N(x,t)$ and $\Delta T(x,t)$ are given by the following diffusion rate equations.

$$\frac{d\delta N(x,t)}{dt} = D\frac{\partial^2 \delta N(x,t)}{\partial t} - (k_{\rm rad} + k_{\rm non})\,\delta N(x,t) \tag{1.6}$$

$$\frac{d\delta T(x,t)}{dt} = k_{\rm non}\frac{Q\delta N(x,t)}{\rho Cp} + D_{th}\frac{\partial^2 \delta T(x,t)}{\partial t} \tag{1.7}$$

By using the photoluminescence (PL) measurement, we already obtained the PL lifetime for this sample. D and $D_{\rm th}$ are the diffusion coefficients of excitons and the thermal diffusivity, respectively. ρ and $C_{\rm p}$ are the density and the heat capacity of materials. Q is the generated heat from unit density of carriers. By solving these equations with $d\delta N(t=0) = N_0$ and $d\delta T(t=0) = 0$, the following relationship is given

$$\delta\widehat{N}(q,t) = N_0 \exp\left[-Dq^2 t - (k_{\rm rad} + k_{\rm non})\,t\right] \tag{1.8}$$

$$\begin{aligned}\delta\widehat{T}(q,t) = &\frac{k_{\rm non}}{k_{\rm rad} + k_{\rm non} + Dq^2}\frac{QN_0}{\rho C_p}\times \\ &\left\{\exp\left[-Dq^2 t - (k_{\rm rad} + k_{\rm non})\,t\right] + \exp\left(-D_{\rm th}q^2 t\right)\right\}\end{aligned} \tag{1.9}$$

where $\delta\widehat{n}(q,t)$ is the Fourier component of $\delta n(x,t)$. q is the grating constant of modulation given by the fringe space of grating (Λ) as $q^2 = 2\pi/\Lambda$. Λ is controlled by the crossing angle of two pump beam as (1.3). Then, the time profiles of the $I_{\rm TG}$ should be fitted by using the following double-exponential decay function with 4 parameter, because ITG is proportional to Δn^2.

$$I_{TG}(t)^{\frac{1}{2}} = \left|A_{\rm ex}\exp\left[-\left(Dq^2 + k_{\rm PL}\right)t\right] - A_{\rm th}\exp\left(-D_{\rm th}q^2 t\right)\right| \tag{1.10}$$

where $A_{\rm ex}$ and $A_{\rm th}$ are the proportional constant of the exciton density grating and the thermal grating, respectively. In the time scale of Fig. 1.7a, the TG signal can be fitted by the single-exponential function because $Dq^2 >> k_{\rm PL}$ and $A_{\rm ex} >> A_{\rm th}$. Obtained D values were plotted against the emission wavelength in Fig. 1.7b. It was found that the D becomes faster with increasing emission wavelength (In-composition). The fluctuation of In-composition leads to the localization centers of excitons and accordingly, emission efficiency becomes high. However, when In-composition becomes too large, the PL intensities and the IQEs become smaller as shown in Fig. 1.4. The turning point is about ~460 nm (blue) for wavelength. We found that D_s of excitons decreases slightly from 0.61 at 460 nm to 0.52 at 470 nm and then increases again with wavelength above 470 nm as shown in Fig. 1.5a. Radiative lifetimes also increase with wavelength (Fig. 1.4). This fact suggests that the localization of exciton becomes weaker with increasing In-composition in this region. The phase separation in the crystal would be enhanced with increasing In-composition as shown in the microscopic emission images in Fig. 1.5. Moreover, In-compositional fluctuation with the submicron scale induces piezoelectric field and causes the exciton acceleration in the lateral direction.

Accordingly, we conclude that the origin of the reduced emission is not re-activation of the NRCs, but inhibition of the radiative processes of the delocalized excitons by the microscopic homogeneity of In-composition and/or exciton acceleration in the lateral direction by the induced piezoelectric field. To develop the emission efficiency, a lot of effort had been done to remove the NRCs (dislocation, defect, or impurity) long time; however, we proposed that the fabrication of the exciton localized center like the quantum dots in the InGaN active layer should be more effective because the origin of the low emission efficiency is not the re-activation of the NRCs but the delocalization of excitons.

Figure 1.8a shows the time profiles of the TG signals with nanosecond pulsed laser under =17 μm, 12 μm, and 8 μm taken for the GaN. For the measurements with nanosecond time scale, a frequency-tripled Nd:YAG laser (355 nm) and a cw He-Ne laser (633 nm) were used as the pump and probe beams, respectively [25]. In this case, the TG signal were fitted by the double-exponential function as shown in (1.10) with $Dq^2 << k_{\mathrm{PL}}$. We assigned the fast and slow decay components as the signals due to the exciton density grating and the thermal grating, respectively. In order to make sure our assignment of these two components, q dependence of the TG signal was measured by changing the crossing angle of the pump beams. All signals could be well fitted by using (1.10). With increasing of q^2, obtained first decay component was not changed, but the slow decay component became shorter because $Dq^2 << k_{\mathrm{PL}}$ in this condition. Moreover, $\tau_{\mathrm{PL}} = 1/k_{\mathrm{PL}}$ of bulk GaN (50 ps) was much shorter than the excitation pulse width (10 ns). Under the wide fringe grating condition, the exciton recombination rates are much faster than the relaxation of the exciton grating by the exciton diffusion. Thus, the thermal grating signal is well generated. On the other hand, under the grating with narrow ($Dq^2 >> k_{\mathrm{PL}}$), a large amount of the exciton density grating is relaxed by the exciton diffusion before the thermal grating is generated. This fact suggests that the thermal dynamics and the exciton diffusion dynamics can be observed selectively by the TG method by controlling the crossing angle of pump beams.

Figure 1.8b shows the relationship between TG decay rate and q^2. This plot shows a good linear relationship with no intercepts at the ordinate. D_{th} of GaN was given by the slope of this plot as 0.44 $\mathrm{cm^2 s^{-1}}$. Thermal conductivity (κ) of GaN can be obtained by the following relationship with the density ($\rho = 6.095$ $\mathrm{gcm^{-3}}$), heat capacity ($C_{\mathrm{p}} = 9.745$ $\mathrm{cal \cdot mol^{-1} K^{-1}}$), and obtained D_{th} values.

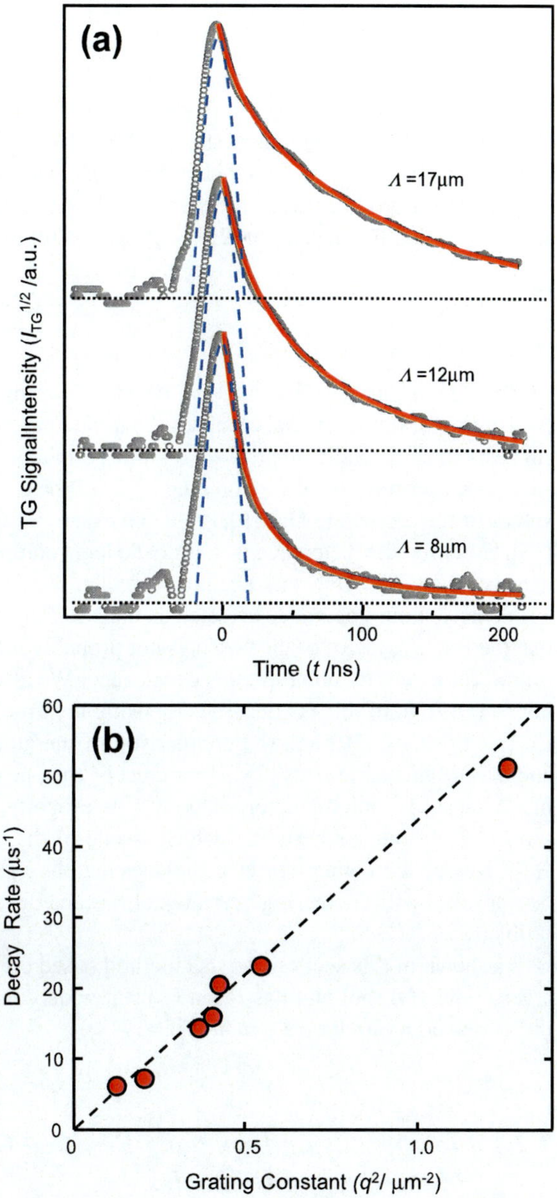

Fig. 1.8 **a** Time profiles of the TG signals under L=17 μm, 12 μm, and 8 μm taken for the GaN. **b** Relationship between the decay rate constant of thermal grating signals and the square of the grating constant. The slope is the thermal diffusivity

$$\kappa = D_{th} \rho C_p \tag{1.11}$$

The obtained κ of bulk GaN in this study was 1.3 W $\cdot$ cm^{-1}K^{-1}. The κ values of GaN have been measured and reported by many groups and known to vary from 1.3 to 2.2 W $\cdot$ cm^{-1}K^{-1} depending on crystal quality [33–37]. Because the TD density of the GaN sample used this study is $\sim 10^{10}$cm^{-2}, obtained κ values should be reasonable.

τ_{non} of excitons is given by the thermal velocity of excitons (V_{th}), the capture cross section (σ), and the density of NRCs (N_{NRC}) as follows:

$$\tau_{non} = \frac{1}{N_{NRC}\ V_{th}\ \sigma} \tag{1.12}$$

The TD density of GaN ($10^9 \sim 10^{10}$cm^{-2}) is much higher than that of ZnSe ($\sim 10^6$cm^{-2}); however, GaN has large η_{int} value as large as that of ZnSe. This fact suggests that V_{th} and/or σ in GaN should be much smaller than that in ZnSe. V_{th} should be related to the diffusion length (L) of the carriers. Actually, the reported L values of the carriers in GaN (0.1 $\sim$ 1.0 μm) are shorter than that in ZnSe (10 μm) [38]. However, the difference in L is not so large compared with the difference in TD density. Then, σ in GaN must be much smaller than that in ZnSe. This suggests that TD density should be not so affected as the NRC in GaN. Therefore, we considered that the limiting factor of the NRCs is not probably macroscopic defects but another origin, for example, point defects or vacancy. We also reported that τ_{non} of GaN is not so sensitive to the TD densities by using the time and spatially resolved micro-PL spectroscopy [39] and the comparison of the atomic force microscope (AFM) and SNOM measurements [28]. This must be the one of most remarkable properties of GaN-based semiconductors. Due to this property, GaN has high light-emitting properties despite its high TD, which is one of the most important advantages of GaN. However, in other words, it suggests that the light-emitting properties may not be improved with increasing crystal qualities and decreasing TD, which is a serious problem.

As shown in this section, the TG method based on third-order nonlinear optics is a very powerful tool and has recently been widely used to observe similar exciton diffusion and nonradiative dynamics [40–42].

1.2.4 Temporally and Spatially Resolved Observations of the Exciton Dynamics

We succeeded in the direct observation of the exciton diffusion, nonradiative recombination, and thermarization processes by using the TG method. The actual correlation between the nonradiative process and the TD is still unknown because the detailed observations of the nonradiative process have been very difficult. Therefore, we tried to perform the temporally and spatially resolved observations of the

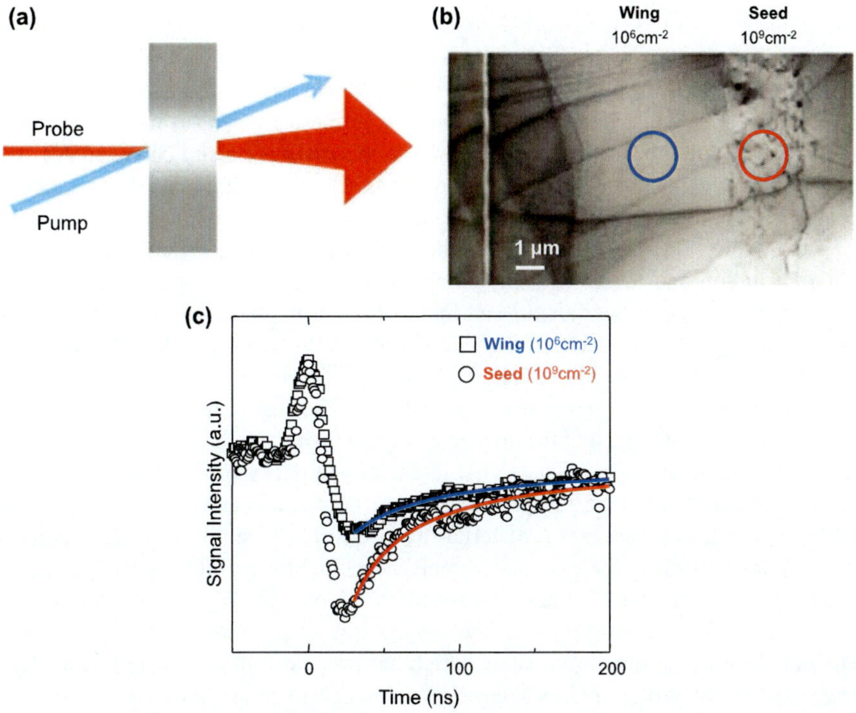

Fig. 1.9 **a** Configuration of the pump beam and probe beam of the transient lens (TL) method. **b** SEM image of the air-bridged lateral epitaxial grown (ABLEG) GaN. **c** Time profiles of the suite selective microscopic transient lens taken for ABLEG-GaN at seed and wing regions

nonradiative exciton dynamics by developing the time-resolved transient lens (TL) technique [43]. Figure 1.9a shows the configuration of the pump beam and probe beam of the TL method. TL method is the third-order nonlinear spectroscopy similar to the TG method, but only one pump beam is used. Therefore, the signal sensitivity and S/N ratio are much lower than those of the TG method; however, microscopic measurement with spatial resolution is available by focusing the pump beam.

Figure 1.9b shows the image of the scanning electron microscope (SEM) taken for the air-bridged lateral epitaxial grown (ABLEG) GaN sample. ABLEG-GaN has two regions, one is a seed region having high TD ($\sim 10^9 cm^{-2}$) and another is a wing region having low TD ($\sim 10^6 cm^{-2}$). We measured the optical properties for each region and compared each other; however, remarkable enhancement of emission efficiencies and lifetime of low dislocated GaN and InGaN has so far not been found. Here, we compare the nonradiative dynamics for each region by the time-resolved micro TL method.

Figure 1.9c shows the time profile of the micro TL with nanosecond pulsed laser taken for ABLEG-GaN [43]. The experimental setup is similar to the TG method with a nanosecond pulse laser. A frequency-tripled beam of a Nd:YAG laser (355 nm) and a He-Ne laser (633 nm) were used as pump and probe beams, respectively. Both pump and probe beams were focused on a sample by an objective lens ($\times 100$). The spot sizes of both beams were about 3 μm in diameter. By the excitation with a Gaussian spatially distributed pump beam, the refractive index (n) in the sample is also spatially modulated by the modulation of the exciton density change (δN) and the temperature change (δT). Such modulations act as the TL. Focus/defocus of the probe beam at the TL was detected by a photomultiplier tube with an optical fiber.

The obtained TL signal has two components, the spike-like component just after photoexcitation and the slow decay component. The fast component and the slow component are due to the defocus and focus of the probe beam at the transient concave and convex lens effects in the sample, respectively. From $(\partial n/\partial N) > 0$ and $(\partial n/\partial T) < 0$, we assigned the fast and slow components to the modulations of the exciton density and the temperature, respectively. The signal intensity and decay of the fast component represent the created exciton density and the exciton recombination/diffusion, respectively. The signal intensity and decay of the slow component represent the amount of the generated heat by the nonradiative recombination process and the thermal conductivity of materials, respectively. By solving the rate equation of (1.6) and (1.7) with the cylindrical coordinates, the time profile of the TL signal can be analyzed and the κ values were obtained 2.2 $\mathrm{W \cdot cm^{-1}K^{-1}}$ at seed region and 2.3 $\mathrm{W \cdot cm^{-1}K^{-1}}$ at wing region. The κ values of each region were not so different and close to the maximum value which has ever been measured [37]. This fact suggests that the ABLEG-GaN should have the highest crystallinity.

Although we could not find the difference in the values at each region, we found a remarkable difference in the thermal signal intensity at each region. We could compare the thermal signal intensities by the nonradiative processes of unit exciton densities at each region by normalization of the exciton signal intensities. It was found that the heat generation at the wing region was much smaller than that at the seed region. At the wing region, low TD should reduce the nonradiative processes of excitons. A similar result was obtained for the InGaN/GaN 3QW structure grown on ABLEG-GaN [43]. As mentioned above, we knew that the efficiency of emission of each region is the almost same at room temperature [39]. We considered that the nonradiative processes of excitons should not be a dominant pathway of thermalization in GaN and InGaN. This is the possible reason that TD is not effective for the emission efficiency of GaN and InGaN/GaN though it acts as the NRCs.

As mentioned above, the IQE values of GaN-based materials have been usually estimated from the temperature dependence of the PL intensities by assuming η_{int}=100% at the quite low temperature. Recently, it was reported that highly defective and nonradiative processes can be still active even at a low temperature so we must be careful to estimate the η_{int} values by this method [44]. In recent years, some studies have been reported to estimate accurate η_{int} values by using an integrating sphere [45] and the photoacoustic spectroscopy [46]. Figure 1.10 shows the temperature dependence of the PL and TL intensities taken for bulk GaN. By using the TL method, we can observe the radiative and nonradiative processes at the same time as shown in this figure. Both dependencies behave similarly, but are not completely complementary to each other. In order to estimate the η_{int} value more accurately,

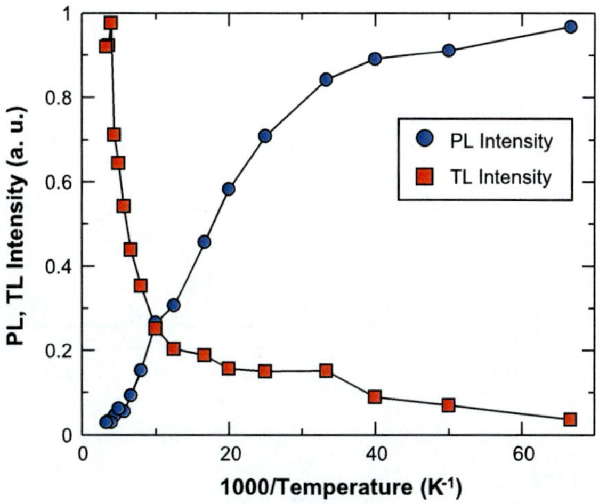

Fig. 1.10 Temperature dependence of the PL and TL intensities taken for bulk GaN

it must be important to compare and analyze the behavior of both radiative and nonradiative processes [47].

In order to perform the direct observation of the NRCs with nanoscale spatial resolution, we combine the TL measurement to the scanning near-field optical microscopy (SNOM) [48, 49]. The experimental setup of the SNOM-TL is shown in Fig. 1.11a. We used a time-modulated 400 Hz InGaN laser 405 nm and a cw-AlGaAs laser 830 nm as pump and probe beams, respectively. Both beams were coupled into a metal-coated optical fiber tip. The aperture size at the end of the tip was 200 nm, and the pump and probe beam power levels, before coupling into the optical fiber, were 5 and 1 mW, respectively. The NSOM-PL signal and the NSOM-TL signal were separated by a dichroic mirror and colored glass filter and detected with a photomultiplier and a highly sensitive femto-watt InGaAs photodetector with a lock-in amplifier system, respectively. The distance between the tip and the sample surface is controlled to 10 nm by shear-force detection. The sample stage was scanned in an x-y direction by a piezo actuator.

Figure 1.11b shows obtained NSOM-PL and NSOM-TL images [49]. We found a submicron-scale spatial inhomogeneity in both NSOM-PL and NSOM-TL images and a clear correlation with each other. The bright regions in NSOM-PL, surrounded by the light blue lines, are corresponding to the bright regions in NSOM-PL. These regions denote the areas which have high recombination center density. Interesting regions are the dark regions in NSOM-PL surrounded by the orange lines. These regions are corresponding to the bright regions in NSOM-PL. In these regions, the exciton density is high but the emission is low. We consider that these regions denote the NRCs, which had not so far been detected directly. The NRCs should be attributed to the point defects and/or threading dislocations, which are present all over the sample but are too small to detect by this technique. Their densities are distributed inhomogeneously and the high-density areas are detected as nonradiative recombination

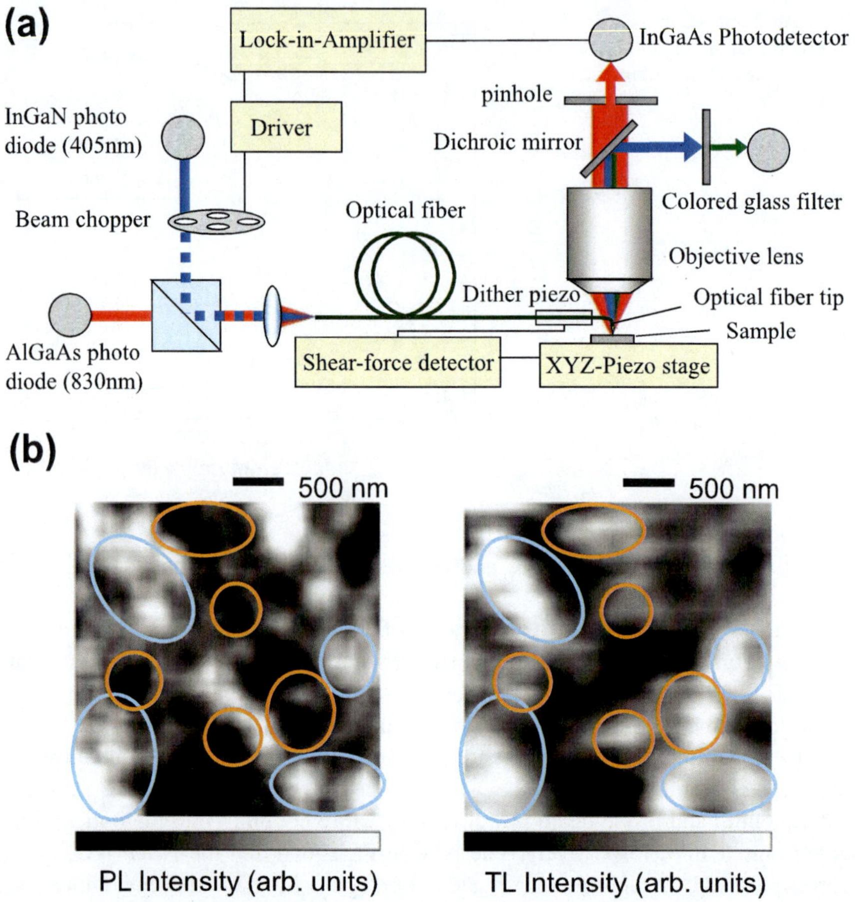

Fig. 1.11 **a** Experimental setup for a scanning near-field optical microscopic transient lens (SNOM-TL). **b** SNOM-PL and SNOM-TL images for InGaN/GaN QW. Adapted from [49] Copyright© 2005, AIP Journals

centers. We conclude that the SNOM-TL are powerful techniques to observe the nonradiative carrier dynamics, which has been very difficult to detect by the conventional methods, in semiconductor materials. The SNOM-TL image gives the exciton density distribution in the active layer. By comparing SNOM-PL and SNOM-TL images, both the radiative and nonradiative recombination centers were unambiguously discriminated with submicrometer scales. This method we presented was also highlighted in Science [Science, **310**, 591, (2005)].

1.3 Controlling the Exciton Dynamics and Light Emissions Based on Nanophotonics and Plasmonics

1.3.1 Emission Enhancement by Fabricated Nanostructures

Based on the direct observations of the radiative and nonradiative exciton dynamics, we propose that controlled spatial inhomogeneity in InGaN active layer should be more effective than making the crystal quality higher by reducing the TD density in order to improve the emission efficiency of InGaN LEDs with high In-composition. For example, making nano-structure may be one of the promising methods to control the spatial inhomogeneity in InGaN and improve the emission efficiencies by increasing k_{rad}. In this section, we show the enhancements of PL of InGaN/GaN by making artificial nanostructures.

We fabricated the arrayed nano-pillar structures of InGaN by using the electron beam lithography and dry etching process. Figure 1.12 shows schematic diagram (a) and SEM image (b) of the nano-pillar array structures [24, 50]. The patterns were defined into a 200-nm-thick polymethylmethacrylate (PMMA) resist spun onto the samples and exposed in a field emission electron beam lithography system. After lithography, the beam patterns were transferred into the samples by the dry etching process. For the dry etching process, we used the chemically assisted ion beam etching (CAIBE). CAIBE is characterized by employing both chemical and physical etching processes, so that anisotropic but fast etching is possible under mild conditions [51]. In this study, Xe was used for sputtering, while Cl_2 was used for chemical etching. By this method, we succeeded to fabricate nano-pillars whose smallest size was down to 50 nm. Figure 1.12 also shows the microscopic PL images of (c) un-patterned and (d) nano-structured InGaN/GaN sample. Both images were taken by the same conditions and shown in the same contrast, so it is clear that the emission intensity of the nano-pillar arrays is much stronger than that of the un-patterned sample. We also found that both the IQE and the emission rate of InGaN/GaN are increased by making the nano-pillar array structures [50]. Moreover, we found an unexpected interesting phenomenon that is the remarkable blinking of emission from the InGaN nano-pillar arrays. This very impressive phenomenon of blinking has been reported for CdSe-based quantum dots structures [52] and is expected to be applicable to optical memory storage circuits and computing. We found very remarkable blinking phenomenon for InGaN/GaN nano-pillars though this phenomenon was rarely reported for InGaN/GaN materials [53]. The nano sizes and arranged structures of light-emitting materials may enhance the blinking but the detailed mechanism is still not clear.

The origin of the intense emission should be due to the strong confinement of the excitons into the nano-pillar structures. This strong confinement should increase the electron–hole wave function overlap and radiative recombination rates, while the exciton localization is no longer effective in the as-grown sample. Moreover, the small-sized In-rich regions of the nano-pillar structures may reduce the QCSE. The internal electric field should be reduced by the partial strain relaxation promoted by

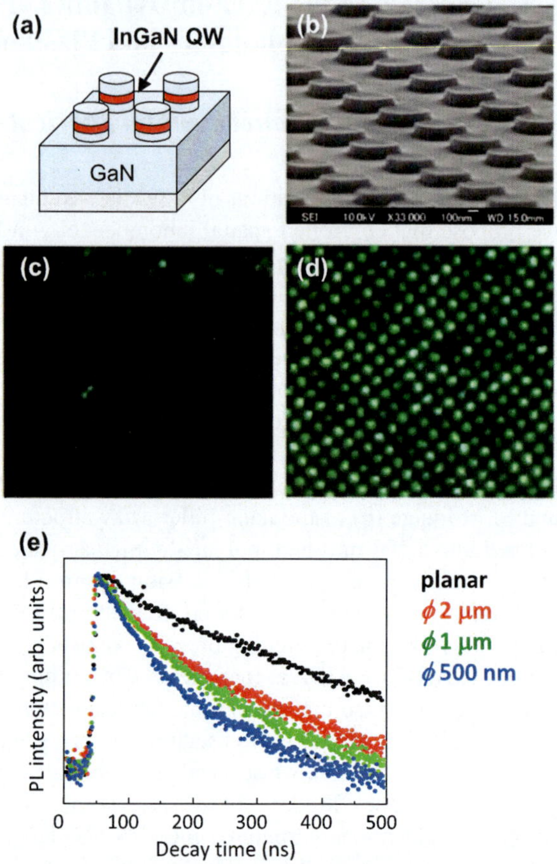

Fig. 1.12 Schematic diagram (**a**) and SEM image (**b**) of the nano-pillar array structures of InGaN/GaN fabricated by E-beam lithography and dry etching. Microscopic PL images of non-structured (**c**) and nano-pillar array with 500 nm diameter (**d**). **e** PL decay curves acquired at the PL peaks for planar QW and nano-pillars at 4.2 K. Adapted from [49] Copyright© 2010, AIP Journals

the size reduction. The enhancement of the exciton localization and the reduction of the QCSE increase the radiative recombination rates and the IQEs. The small structures also can enhance the light extraction efficiencies.

Fabricated nanostructures can control exciton dynamics and enhance the emission rate. Moreover, emission rates can be increased by the light confinement into nanostructures. When the exciton is located in a cavity resonator, the spontaneous emission rate of the exciton is proportional to the quality factor (Q) and inversely proportional to the mode volume (V) of the cavity resonator. This phenomenon is known as Purcell effect and various types of cavity resonators have been fabricated to increase Q and reduce V values [54]. For example, photonic crystals have attracted attention as a promising platform for the realization of compact and efficient nanocavities and lasers. Figure 1.13a shows the SEM image of the photonic crystal nanocavity with InGaAsP QW [55]. The QW structure includes 9 nm-thick InGaAsP active layers (1550 nm) and 20 nm-thick InGaAsP barrier layers (1220 nm) on a 330 nm-thick InGaAsP slab. The photonic crystal structure is a free-standing membrane patterned

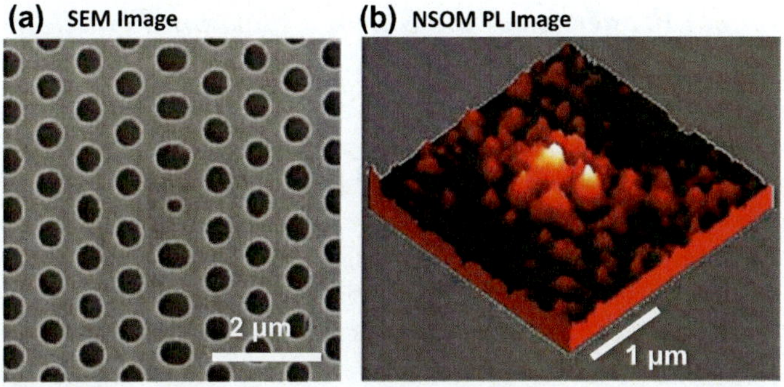

Fig. 1.13 **a** SEM image of the photonic crystal nanocavity with InGaAsP QW. **b** Near-field image of the cavity mode of the photonic crystal nanocavity

with a triangular lattice of holes within which cavity based on fractional edge dislocation is defined. Figure 1.13b shows the near-field image of the photonic crystal nanocavity [56]. The experimental setup of the SNOM is similar to Fig. 1.11a. A 20 ns pulsed diode laser (780 nm) was used as the excitation laser source with the collection mode (C-mode) measurements of the SNOM. The PL signals were distinguished from the excitation laser by using the colored glass filter with a cut-off wavelength of 850 nm and detected with a femto-watt InGaAs photodetector. The metal-coated fiber tip, with a small aperture size, enables us to distinguish between localized cavity modes and propagating far-field modes and to obtain more precise mode profiles when the tip probes into holes of photonic crystals. The best resolution in our system is as small as 50 nm. This is the first report to observe the clear optical mode images obtained by SNOM on very small photonic crystal cavities. The Q and V values of the nanocavity were measured directly as $\sim 10^4$ and 0.03 μm^3. We have confirmed that the photonic crystal structure results in significantly higher emission rates and also stimulated emission with a low threshold as nano-laser. However, InGaN/GaN has been very difficult to fabricate photonic crystal structures because much smaller structures were required.

We concluded that designing and fabrication of nanostructures should be a promising method to improve the optical properties. However, super-bright LEDs as an alternative to conventional fluorescent light bulbs have not so far been achieved in spite of a lot of efforts about these nanostructures. The problem may be due to the difficulty and expense of nano-fabrication techniques. Thus, the development of a new method, which must be much easier and more effective, has been earnestly desired for further improvements of emission efficiencies of InGaN LEDs. In the next section, we propose another method based on "plasmonics" as one solution to dramatically increase the efficiencies of light-emitting materials and devices.

1.3.2 Optical Properties of the Surface Plasmon Polariton

Plasmonics enable the manipulation of light waves at the nanoscale–that is, at a much smaller scale than the wavelength scale–by inducing resonance with surface plasmons (SP) generated at the metal/dielectric interface [57]. Following the remarkable developments of nanotechnologies in recent years, plasmonics has been expected to become a fundamental and platform technology that will lead to various new classes of nano-optical applications. Various new optical nanotechnologies operating below the wavelength size have been proposed, including plasmonic bandgap structures [58], plasmonic waveguides [59], and anomalous light transmission using nanohole arrays [60]. In 2001, our project team at the California Institute of Technology in collaboration with Atwater have introduced the term "plasmonics" for new research fields dealing with light and its interactions at the nanoscale via SP [61]. Plasmonics has been expected to lead to the development of various new fundamental technologies, such as nano-optical circuits and computing, that will support the future ultrasmart society [62].

Future applications for plasmonics include the development of high-efficiency light-emitting diodes (LEDs). It was theoretically proposed by the group led by Yablonovitch, the inventor of photonic crystals, that plasmonics should be useful for increasing the emission efficiencies of materials [63, 64]. Based on this theory, in 2004, we presented the first report of large photoluminescence (PL) enhancements for InGaN/GaN quantum well (QW) materials coated with Ag thin films [65]. The usefulness of plasmonics for increasing the emission efficiencies of materials and devices is now well known, and this paper [65] was the first experimental report demonstrating this utility.

Figure 1.14a shows the electric-field distribution of the propagating modes of the surface plasmon polariton (SPP) generated at metal/dielectric interface calculated by the finite-difference time-domain (FDTD) calculations [24, 66]. The electric fields are strongly confined near the metal surface, thereby endowing materials near the metal interface with novel optical properties and functions. An SPP can interact with a light wave because it includes a transverse wave of charge fluctuations at the metal surface, whereas the compressional wave of an SP consists only of a longitudinal wave. The wave vector (k_{SP}) of an SPP propagating parallel to such an interface can be written as follows [67]:

$$k_{SP}(\omega) = \frac{\omega}{c}\sqrt{\frac{\varepsilon_m(\omega)\varepsilon_d(\omega)}{\varepsilon_m(\omega) + \varepsilon_d(\omega)}} \tag{1.13}$$

where the relative permittivities of the metal and the dielectric material are denoted by ε_m and ε_d, respectively, and ω and c are the angular frequency and velocity in vacuum, respectively, of the light.

The dispersion diagram, which shows the relationship between k_{SP} and angular frequency (ω), can be obtained from (1.12), and Fig. 1.14b presents the dispersion diagrams for Al/GaN, Ag/GaN, and Au/GaN interfaces. To generate this figure, the

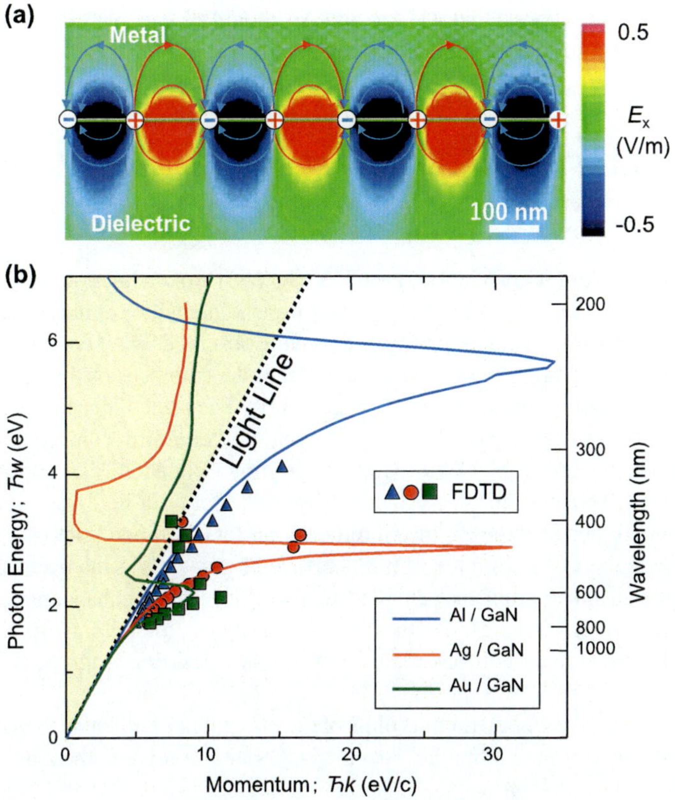

Fig. 1.14 **a** Electrical field distribution of the surface plasmon polariton calculated by the FDTD simulations. **b** Dispersion diagram of the propagation mode of the SPP at Al/GaN, Ag/GaN, and Au/GaN interfaces

relative permittivity of GaN was set to 6 without wavelength dispersion. and can be converted into the photon energy ($\hbar\omega$) and the momentum ($\hbar k_{SP}$), respectively, through multiplication by the reduced Planck constant ($\hbar$). The black dotted line represents the light line defined by $k_d = \sqrt{\varepsilon_d}\omega/c$, where k_d is the wave vector of a light wave propagating in the dielectric material, which can exist only inside the area defined by this line, as shown by the shading in Fig. 1.14b. In contrast, SPP dispersion exists outside of the light line because k_{SP} can be larger than k_d when $-1 < \varepsilon_d + \varepsilon_m < 0$. This is the most important property of SPPs; consequently, a metal with a negative permittivity is necessary to generate SPPs. A large k_{SP} means that the wavelength is compressed and the velocity is slowed down at the metal surface. The value of ω at the maximum value of k_{SP} is called the SP resonance frequency (ω_{SP}). The SP resonance energies (wavelengths) are 5.7 eV (220 nm), 2.9 eV (430 nm), and 2.3 eV (540 nm) for Al/GaN, Ag/GaN, and Au/GaN interfaces, respectively. This suggests that Al, Ag, and Au should be suitable for use in the UV, blue, and red regions of the spectrum, respectively.

1.3.3 Mechanism of the Surface Plasmon-Enhanced Light Emissions

PL spectra taken for InGaN/GaN single QWs (SQWs) (broken lines) and 3QWs (solid lines) were shown in Fig. 1.15 [65, 68]. The PL peak intensities of uncoated samples were normalized to 1 and huge enhancements were observed by Ag coating especially at the shorter wavelength region. The wavelength dependences of the enhanced PL intensities were almost the same for SQW and 3QW. The inset of this figure shows the sample structure of InGaN/GaN 3QW and excitation/emission configuration of PL measurement. InGaN/GaN, QWs (3 nm) were grown via MOCVD, and Ag layers (50 nm) were deposited on the surfaces of these wafers via high-vacuum thermal evaporation. To perform the PL measurements, an InGaN diode laser (406 nm) was used to excite the QWs from the bottom surface of the wafer. This backside access is the important innovation that enabled us to achieve light enhancement through SP coupling for the first time. With this configuration, we could avoid absorption losses at the metal layer and achieve effective light extraction from the SPPs at the interface.

Figure 1.16a shows the Arrhenius plots of the integrated PL intensities from InGaN QWs separated from Ag films by 10 nm spacers and compares these to uncoated

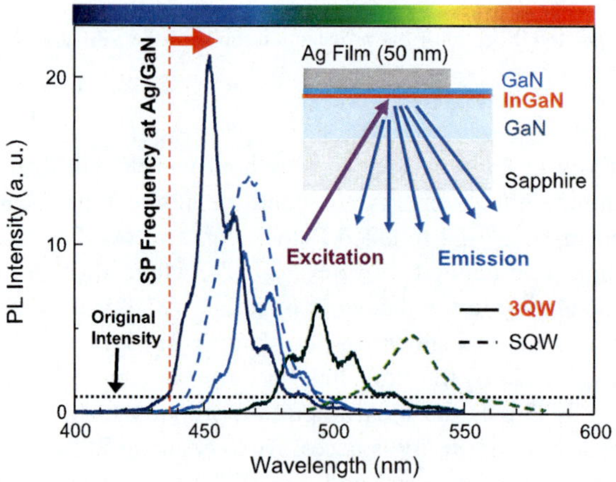

Fig. 1.15 SP-enhanced PL spectra of InGaN/GaN SQW with peak wavelength at 470 and 530 nm (broken lines) and 3QW with peak wavelength at 450, 460, and 500 nm (solid lines) coated with Ag. The PL peak intensity of each uncoated sample was normalized to 1. (Inset) Sample structure of InGaN/GaN QW and excitation/emission configuration of PL measurement. Adapted from [68] Copyright© 2010, Wiley Journals

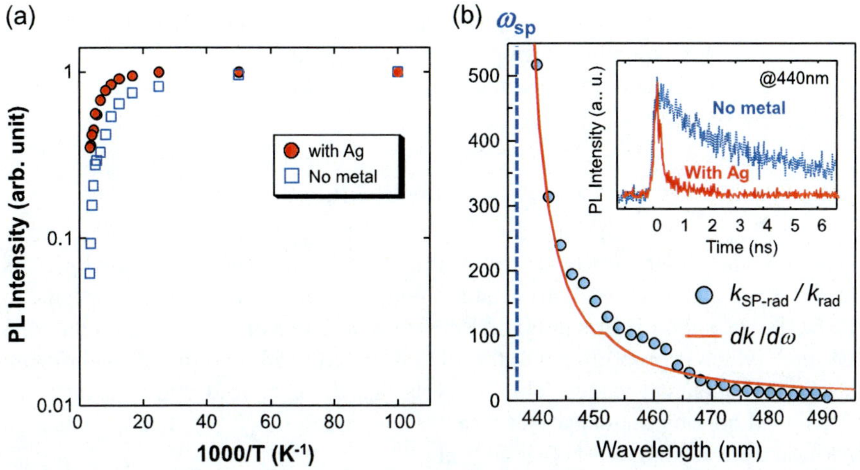

Fig. 1.16 **a** Arrhenius plots of the integrated PL intensities of InGaN/GaN QWs with Ag and uncoated sample with 10 nm GaN spacers. PL integrated intensities at 10 K were normalized to 1. **b** Enhancement factor of the spontaneous emission obtained by the ratios of SP coupling rates and radiative recombination rates. The solid line is $dk/d\omega$ of the SPP at Ag/GaN interface. (Inset) PL decay profiles of uncoated and Ag-coated InGaN/GaN QW at 440nm. Adapted from [24] Copyright© 2009, IEEE Journals and [68] Copyright© 2010, Wiley Journals

samples [24, 65]. The IQE values (η_{int}) from uncoated QWs were estimated as 6% at room temperature by assuming η_{int} ~100% at 10 K. These IQE values increased to 40% after Ag coating, explainable by spontaneous recombination rate enhancements through SP coupling. This suggests that the spontaneous emission rate becomes much faster by the SP coupling. We investigated the direct observation of SP-coupled spontaneous emission rate by using the time-resolved PL measurements [68, 69].

The inset of Fig. 1.16b shows the PL decay profiles of uncoated and Ag-coated InGaN/GaN QW samples emitting at 440 nm. The frequency-doubled output from a mode-locked Ti:Al_2O_3 laser (wavelength: 400 nm, repetition rate: 80 MHz, pulse width: 1.5 ps) was used to excite the InGaN QWs from the bottom surfaces of the wafers. The emission lifetime of the Ag-coated sample (~200 ps) was 30 times faster than that of the uncoated sample (~6 ns). This effect of enhancing the spontaneous emission rate is a kind of Purcell effect [54, 63, 64]. The spontaneous emission rates of excited atoms, molecules, or excitons partially depend on the environment if they are located in optical cavities. With SP coupling, the PL decay rate (k^*_{PL}) should be increased by the increasing of the radiative rate (k^*_{rad}) as $k^*_{\text{PL}} = k^*_{\text{rad}} + k_{\text{non}}$. The Purcell enhancement factor (F_{p}) is defined in terms of the emission enhancement ratio as follows:

$$F_p(\omega) = \frac{k^*_{\text{rad}}(\omega)}{k_{\text{rad}}(\omega)} \tag{1.14}$$

The PL decay time profiles were fitted with single-exponential functions, and the obtained F_p values are plotted against the wavelength in Fig. 1.16b.

Theoretically, the spontaneous emission rate of excitons (=k_{rad}) is given by Fermi's Golden rule, as follows [63, 64]:

$$k_{rad}(\omega) = \frac{2\pi}{\hbar}|\mathbf{d} \cdot \mathbf{E}(\omega)|^2 \rho(\omega) \quad (1.15)$$

where $\mathbf{d}$ is the dipole moment of an exciton, $\mathbf{E}(\omega)$ is the electric field, and $\rho(\omega)$ is the density of states (DOS). This equation suggests that F_p should be proportional to the DOS. The solid red line in Fig. 1.16b shows the slope of the dispersion diagram ($dk/d\omega$), which is proportional to the DOS of the SPP. We find that F_p and $dk/d\omega$ have nearly identical values. This suggests that the enhanced emission rates are determined only by the density of states, as described by (1.14). The F_p values in this case were as high as ~500 at 460 nm.

The enhanced IQE (η^*_{int}) can be estimated by the initial value of IQE (η_{int}) and F_p by the following relationship [66, 70]:

$$\eta^*_{int} = \frac{C'_{ext} F_p \eta_{int}}{(F_p - 1)\eta_{int} + 1} \quad (1.16)$$

where C'_{ext} is the efficiency from SPP to light emission. Figure 1.17 shows the relationship between the enhanced IQE and the Purcell enhancement factor for several values of the initial IQE when ignoring the thermal loss of the SPP (C'_{ext}=1). This situation should be achievable by controlling the size of metal nanograin structure which was self-organized during metal deposition. The sizes of the nanograin can be controlled by adjusting the deposition conditions and used as nano-diffraction grating to extract light emission from the SPPs at the metal interface [71]. This figure suggests that it should be possible to increase the IQE using the SP coupling method

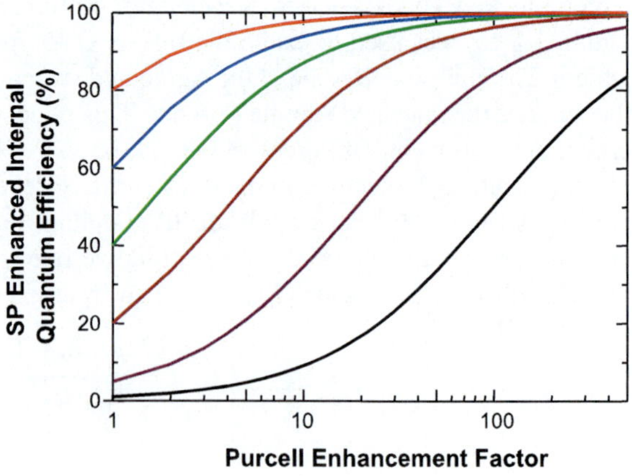

Fig. 1.17 Relationship between the enhanced IQEs and Purcell enhancement factors with several values of the initial IQEs. Adapted from [66] Copyright© 2017, Elsevier Journal

even when the initial IQE value is high. Often, the SP coupling method has been misunderstood to be useful only when the initial efficiency is very poor, not for highly efficient light-emitting materials. However, Fig. 1.17 suggests that this method can be used to increase the IQE from 80% to almost 100%.

SP resonance is also a type of nano-optical resonator, and F_p is given by the quality factors (Q) and mode volumes (V) as follows [54]:

$$F_p = \frac{3}{4\pi^2}\left(\frac{Q}{V/V_\lambda}\right) \tag{1.17}$$

where $V_\lambda = \lambda^3$ is the volume of the light in free space. In recent years, nano-lasers based on the SP resonance with propagating mode [72, 73] or localized mode [74, 75] have been reported and have attracted increased attention. By using plasmonic nanocavities, very small V values much smaller than V_λ are available. On the other hand, Q values of plasmonic nanocavities are usually up to ~10 and difficult to increase. Quite recently, we designed the unique random metal nanohemisphere on mirror (NHoM) structure to tune the SP resonance in a flexible manner [76]. We achieve the quality factor $Q > 200$ that is surprisingly high for the localized SP mode of nanoparticles. Flexible tuning of the SP resonance with the use of NHoM is envisaged to lead to the development of new applications and technologies in the field of plasmonics and nanophotonics.

The most important targets of SP-enhanced light emission are device applications for high-efficiency LEDs. The SP coupling technique increases the IQE of a device by increasing the spontaneous emission rate. This suggests that this method should be applicable in cases involving electrical pumping because the IQE does not depend on the pumping method used. Therefore, we attempted to design super-bright plasmonic LEDs operating under electrical pumping. Recently, a great number of groups have reported SP-enhanced LEDs based on our technique [77, 78]. Quite recently, we attempted to fabricate plasmonic LEDs with very thin p^+-GaN layers based on the structure shown in Fig. 1.18a [79]. We succeeded in fabricating LED devices with p^+-GaN layers with a thickness of 20 nm by controlling the level of doping in the p-GaN layer. We achieved an enhancement of the electroluminescence (EL) using the fabricated plasmonic LED device structure, as shown in Fig. 1.18b. A twofold enhancement was achieved in the cited study; however, we believe that further enhancement should be possible through optimization of the metal and device structures. SP-enhanced LEDs also have several other advantages. For example, SP coupling can allow various effects that are detrimental to the emission efficiency to be avoided, such as the QCSE [80], the leakage through an ohmic contact [81], and the efficiency droop effect [82, 83].

The next important challenge is the extension of the SP resonance phenomenon into the UV and IR wavelength regions to enable a wider range of applications. In particular, deep UV (DUV) LEDs are expected to be useful for diverse applications, such as sterilization, lithography, optical memory, and various biological purposes. In 2010, we succeeded in enhancing DUV emission by increasing the IQEs of

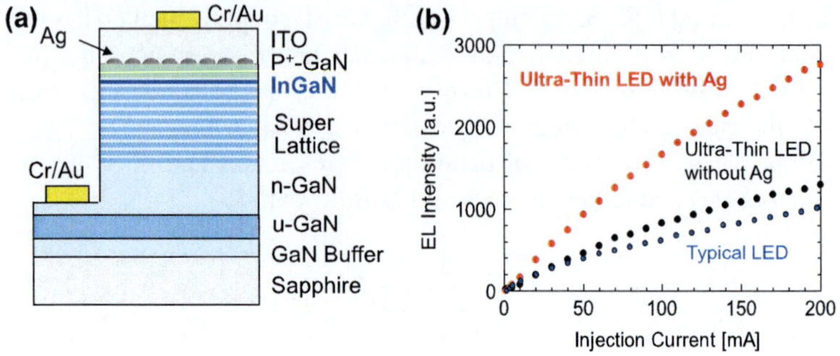

Fig. 1.18 **a** Sample structure of the InGaN-based plasmonic LED by employing the very thin p^+-GaN **b** EL intensities under the several values of the driving currents. Adapted from [79] Copyright© 2017, AIP Journal

AlGaN/AlN QWs by means of the exciton–SP resonance enabled by Al nanograin structures [84]. Similar method was used to enhance near UV emission around 360 nm from AlGaN/GaN-based QWs [85] and LEDs [86]. Geo et al. also reported a twofold enhancement of the DUV (~250 nm) light emission from AlGaN/AlN-based QWs with Al films [87]. In this case, only the light extraction was enhanced by means of the SP resonance enabled by the Al films.

Figure 1.19a shows the PL spectra measured using a picosecond DUV laser (215 nm) [84]. We grew an 5-layer AlGaN (15 nm)/AlN (14 nm) QW structure on an AlN (600 nm) layer on a sapphire substrate and deposited an Al film on top of the wafer. Figure 1.19b shows the size-controlled random nanograin structure observed in atomic force microscope (AFM) taken for the Al surface. We achieved a sevenfold enhancement of the DUV emission near 260 nm. From temperature-dependent measurements of the PL intensities, we found that this remarkable enhancement could be attributed to an increase in the IQE, as shown in Fig. 1.19c. We also found, through TRPL measurements, that the DUV emission rate was increased as shown in Fig. 1.19d.

1.3.4 Enhanced Mechanism and Exciton Dynamics with Submicron to Nanometer Scale

In spite of a lot of efforts and reports for plasmonic LEDs, the enhancement factors have been so far as large as only twice by electrical pumping, though more than 10 fold enhancements were achieved for photo pumping [66, 77, 78]. In order to achieve more enhancement factors of this method, the further optimization of both device and metal structures by deeply understanding the SP coupling mechanism should be necessary. For example, the relationship between the SP-enhanced light emission and exciton dynamics in InGaN/GaN QW is still not so clear. As mentioned above,

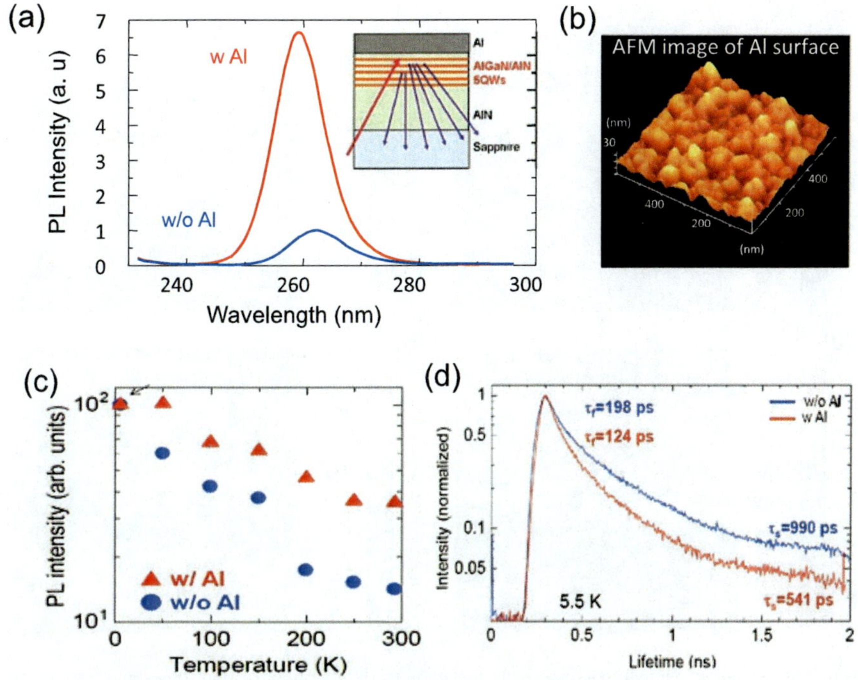

Fig. 1.19 **a** Enhancement of deep UV emission from AlGaN/AlN with/without Al film. (inset) Sample structure of the AlGaN/AlN QWs with Al films. **b** AFM image of the Al nanograin structure on AlGaN/AlN. **c** Temperature dependence of the PL intensities of deep UV emission from AlGaN/AlN with/without Al film. **d** Time profiles of the emission from AlGaN/AlN with/without Al film. Adapted from [66] Copyright© 2017, Elsevier Journal

the IQE of InGaN/GaN QWs strongly depends on the exciton dynamics in active layers, such as exciton localization and quantum confinement Stalk effect (QCSE). Spatial-resolved evaluations of the PL features using microscopic PL measurements are very useful to measure and elucidate such the exciton dynamics which strongly affect the IQEs of emissions as described in Sect. 1.2. In this section, we used a microscope with a moving stage and evaluated the exciton dynamics under the SP resonance to understand its detail mechanism.

PL spectra were measured with a fluorescence microscope using a mercury lamp for excitation (400–410 nm) and a spectrograph (Spectra Pro 2300i) and charge-coupled device (CCD) camera (PIXIS100) [88]. A motorized moving stage was controlled by a stepping motor for scanning with 2-μm step. All of the PL spectra obtained at each pixel were fitted with a Gaussian function, and the values of the PL peak wavelength and intensity were extracted from the Gaussian fitting result.

Figure 1.20 shows the PL mapping results obtained from the blue-emitting sample. Figure 1.20a, b shows the peak intensity and peak wavelength mapping images obtained from the bare region, respectively [89]. Both the PL peak intensity and peak

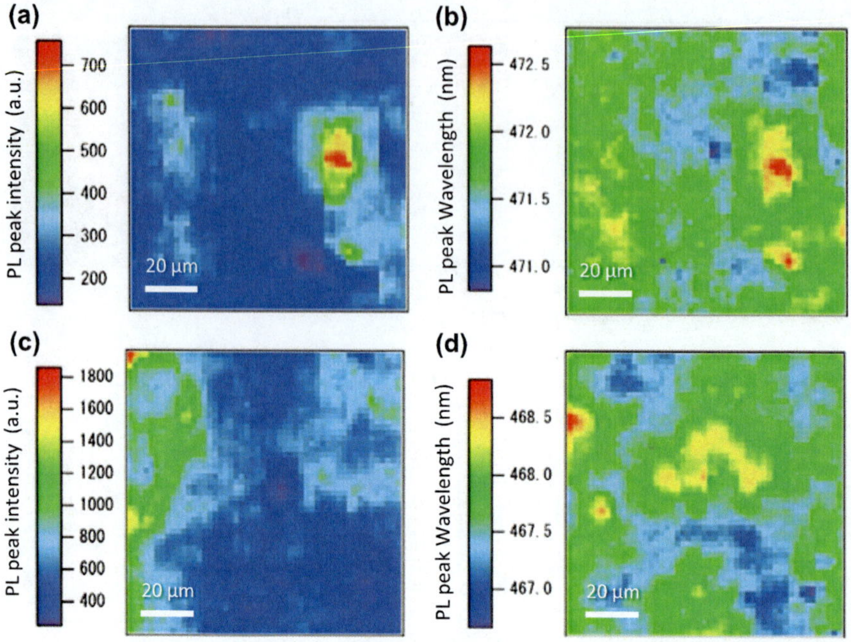

Fig. 1.20 PL peak intensity (**a**) and peak wavelength (**b**) mapping of bare blue-emitting InGaN QWs. PL peak intensity (**c**) and peak wavelength (**d**) mapping of Ag-coated sample. Adapted from [89] Copyright© 2017, AIP Journal

wavelength values exhibited some spatial fluctuation in the scanned area. This figure shows that the regions with higher PL intensities seemed to have longer PL wavelengths. The positive correlation between the PL intensities and peak wavelength is consistent with our previous reports shown in Fig. 1.5. Figure 1.20c, d shows the PL peak intensity and peak wavelength mapping images obtained from the silver-coated region of the same sample. In this region, the PL intensities were enhanced, and positive correlations were no longer observed, which was probably due to the PL enhancement. Similar measurements were performed for the InGaN/GaN with green emission [89].

In order to understand a more detailed mechanism, we plot the relationship between the PL peak intensities and wavelength for blue and green emission in Fig. 1.21 [88, 89]. For the uncoated regions shown in black marks, the regions with higher PL intensities seemed to have longer PL wavelengths for InGaN with blue emissions. This positive correlation is similar to our previous reports (Fig. 1.5) and should be due to the exciton localization effects. The generated excitons diffuse in the QW layer and localized into In-rich regions, which seem as quantum dots structures. The localized excitons are well protected against thermal loss and have high emission efficiencies. For this reason, the region of longer PL wavelength must be much brighter than the shorter wavelength regions in blue emission.

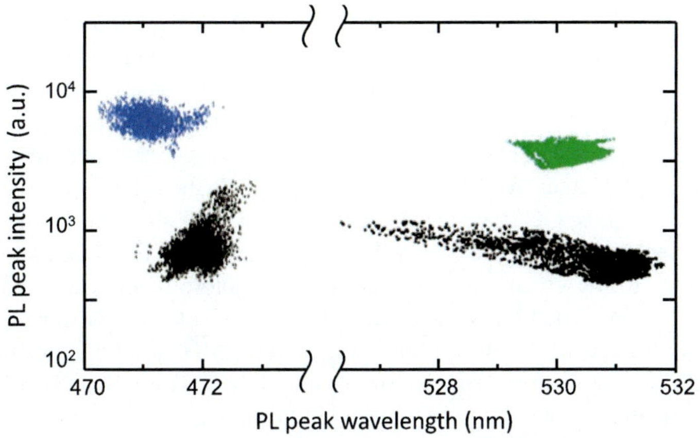

Fig. 1.21 Relationship between the PL peak intensity and the PL wavelength at each pixel for Ag-coated (colored circles) and uncoated (black circles) part of blue- and green-emitting InGaN/GaN QWs. Adapted from [89] Copyright© 2017, AIP Journal

Figure 1.21 shows that the PL peak wavelength was blue shifted and the positive correlation disappeared by Ag coating for blue-emitting InGaN/GaN. This result suggests that the energy transfer from the excitons to the SPs on Ag film is faster than the localization processes. If the SP resonance increases the emission rate faster than the exciton localization, the exciton localization effects should be canceled. Then the positive correlation should be disappeared and the PL wavelength is blue shifted. The exciton localization process is in many cases positive factor to improve the emission efficiencies of the LEDs; however, sometimes negative factor to apply to laser diodes. The canceling feature to the exciton localization of the SP-enhanced emissions would be useful to apply to the low threshold laser diodes in the future.

Opposite to the blue emission, the negative correlation was observed for InGaN QWs with green emissions. Also, the energy distributions of the PL peak wavelength were much broader comparing with the blue emission. We already reported these properties of green-emitting InGaN/GaN QWs as an increase in the number of defects as the nonradiative recombination centers (NRCs) as well as the quantum confined Stark effect (QCSE). It is well known that the QCSE can be screened when the exciton densities were very high after the photoexcitation [90, 91]. After the excitation, with decreasing exciton density, the screening effect was reduced and the QCSE becomes more remarkable. This should be the reason for the very broad distribution of the PL peak wavelength.

Figure 1.21 shows that the negative correlation also disappeared by the SP resonance with Ag coating. One contribution of the SP resonance is similar to the blue emissions, the SP resonance canceled the exciton localization because the SP-enhanced emission rate should be faster than the exciton localization. For the green emission, the exciton localization brings the negative correlation, so the SP resonance can cancel it and also should bring the blue shift of the PL peak wavelength.

Figure 1.21 shows that the main part of the PL peak distribution was blue shifted by Ag coating. However, some parts of the PL peak distribution were red-shifted. The other important process is the influence of the SP resonance on the QCSE. As described above, the SP resonance increases the decay rates of the excitons. The exciton generated in the QW should rapidly decay by the SP resonance; therefore, the exciton density becomes much lower immediately. This should prevent the screening of the QCSE, and the QCSE becomes much stronger. This should bring the red shift of the PL peak wavelength shown in Fig. 1.21. Therefore, the contribution of the SP resonance to the green-emitting InGaN/GaN is more complicated than that of the blue emissions. The SP resonance canceled both the exciton localization and the screening of the QCSE. By these effects, the wavelength distribution of the PL peak wavelength becomes very narrow for the green emission. This also an very important effect to consider the photonic application of plasmonics.

All these effects of the SP resonance with Ag were not observed for the SP resonance with Al because the enhancement mechanisms were quite different [92]. Oppositely to the SP resonance with Ag, the screening of the QCSE can be enhanced by increasingthe exciton density by the light absorption enhancement by the SP resonance with the excitation light at the Al interface [93]. The giant electric field generated at the Al interface by the SP resonance also is able to cancel the QCSE directory. In this study, we found that the SP resonance can control not only the emission rates and efficiencies but also both the exciton localization and the QCSE in the QWs. These exciton dynamics are very important factors to decide the optical properties and functions of the InGaN/GaN QWs. Based on this method, much more flexible controlling and tuning of the optical properties and functions should be available by further optimization of these effects.

1.4 Summary

Temporal and spatial-resolved spectroscopic techniques based on the nonlinear optics and near-field optics are very powerful tools to characterize nonradiative processes in GaN and InGaN/GaN. Such studies are very important to develop the device performances and light-emission properties of optical materials. Based on these observations and characterizations, we proposed a novel method to enhance the emission efficiencies of InGaN/GaN based on nanophotonics and plasmonics. These techniques have been expected to lead to a new class of very-bright and high-speed LEDs and highly efficient solid-state light sources.

Nanophotonics and plasmonics have the potential to dramatically advance conventional optical technologies, such as efficient energy devices, threshold-less nanolasers, high-resolution imaging/fabrication, and high-sensitivity sensors. Moreover, it would lead to the development of future technologies, such as highly integrated nano-optical circuits and nano-optical computing. Such next-generation optical technologies are expected to become one of the fundamental technologies that will support the emerging ultrasmart and internet-of-things societies.

Acknowledgements The author wish to thank Prof. K. Tamada (Kyushu University), Prof. Y. Kawakami and Prof. M. Funato (Kyoto University), Prof. K. Tadatomo and Prof. N. Okada, (Yamaguchi University), and Prof. K. Wada and Prof. T. Matsuyama (Osaka Prefecture University) for valuable discussions and support. This work was supported by the JSPS KAKENHI Promoted Research (20H05622), S (19H05627), and B (18H01903).

References

1. S. Nakamura, T. Mukai, M. Senoh, S. Nagahama, N. Iwasa, Jpn. J. Appl. Phys. **32**, L8 (1993)
2. K. Bando, K. Sakano, J. Light Vis. Environ. **22**, 2 (1998)
3. S. Fujita, Jpn. J. Appl. Phys. **54**, 030101 (2015)
4. Y. Narukawa, J. Narita, T. Sakamoto, K. Deguchi, T. Yamada, T. Mukai, Jpn. J. Appl. Phys. **46**, L963 (2007)
5. Y. Taniyasu, M. Kasu, T. Makimoto, Nature **441**, 325 (2006)
6. A. Khan, K. Balakrishnan, T. Katona, Nature Photonics **2**, 77 (2008)
7. M. Yamada, T. Mitani, Y. Narukawa, S. Shioji, I. Niki, S. Sonobe, K. Deguchi, M. Sano, T. Mukai, Jpn. J. Appl. Phys. **41**, L1431 (2002)
8. J.J. Wierer, M.R. Krames, J.E. Epler, N.F. Gardner, M.G. Craford, J.R. Wendt, J.A. Simmons, M.M. Sigalas, Appl. Phys. Lett. **84**, 3885 (2004)
9. T. Mukai, K. Takekawa, S. Nakamura, Jpn. J. Appl. Phys. **37**, L839 (1998)
10. T. Mukai, and S. Nakamura: Jpn. J. Appl. Phys.**38**, 5735- (1999)
11. K. Okamoto, A. Kaneta, Y. Kawakami, S. Fujita, J. Choi, M. Terazima, T. Mukai, J. Appl. Phys. **98**, 064503 (2005)
12. K. Okamoto, S. Saijou, Y. Kawakami, S. Fujita, M. Terazima, G. Shimomiya, T. Mukai, Proc. SPIE **4278**, 150 (2001)
13. S. Chichibu, T. Azuhata, T. Sota, S. Nakamura, Appl. Phys. Lett. **69**, 4188 (1996)
14. Y. Narukawa, Y. Kawakami, M. Funato, S. Fujita, S. Fujita, S. Nakamura, Appl. Phys. Lett. **70**, 981 (1997)
15. A. Kaneta, K. Okamoto, Y. Kawakami, S. Fujita, G. Marutsuki, Y. Narukawa, T. Mukai, Appl. Phys. Lett. **81**, 4353 (2002)
16. H.J. Jeong, H.MOh. Jeong, C.H. Hong, E.K. Suh, G. , Lerondel, and M. S. Jeong: Sci. Rep. **5**, 9573 (2015)
17. T. Takeuchi, S. Sota, M. Katsuragawa, M. Komori, H. Takeuchi, H. Amano, I. Akasaki, Jpn. J. Appl. Phys. **36**, L382 (1997)
18. P. Walterelt, O. Brandt, A. Trampert, H.T. Grahn, J. Menniger, M. Ramsteiner, M. Reiche, K.H. Ploog, Nature **406**, 865 (2000)
19. K. Nishizuka, M. Funato, Y. Kawakami, Sg. Fujita, Y. Narukawa, and T. Mukai: Appl. Phys. Lett.**85**, 3122 (2004)
20. T. Sugahara, M. Hao, T. Wang, D. Nakagawa, Y. Naoi, K. Nishino, S. Sakai, Jpn. J. Apple. Phys. **37**, L1195 (1998)
21. S. Chichibu, K. Wada, S. Nakamura, Appl. Phys. Lett. **71**, 2346 (1997)
22. K. Okamoto, J. Choi, Y. Kawakami, M. Terazima, T. Mukai, S. Fujita, Jpn. J. Appl. Phys. **48**, 839 (2004)
23. A. Kaneta, M. Funato, Y. Kawakami, Phys. Rev. B **78**, 125317 (2008)
24. K. Okamoto, Y. Kawakami, IEEE J. Sel. Top. Quantum Electron **15**, 1199 (2009)
25. K. Okamoto, Y. Kawakami, Sg. Fujita, and M. Terazima: Anal. Sci. **17**, s312 (2001)
26. K. Okamoto, A. Kaneta, K. Inoue, Y. Kawakami, M. Terazima, T. Mukai, G. Shinomiya, S. Fujita, Status Solidi B **228**, 81 (2001)
27. H. Haag, P. Gilliot, R. Levy, B. Honerlage, O. Briot, S. Ruffenach-Clur, R.L. Aulombard, Appl. Phys. Lett. **74**, 1464 (1999)
28. S. Pau, J. Kuhl, F. Scholz, V. Haerle, M.A. Khan, C.J. Sun, Appl. Phys. Lett. **72**, 557 (1998)

29. B. Taheri, J. Hays, J.J. Song, B. Goldenberg, Appl. Phys. Lett. **68**, 587 (1996)
30. T. Aoki, G. Mohs, M. Kuwata-Gonokami, A.A. Yamaguchi, Phys. Rev. Lett **82**, 3108 (1999)
31. H. Haag, B. Honerlage, O. Briot, R.L. Aulombard, Phys. Rev. B **82**, 11624 (1999)
32. H. Kogelmik, Bell. Syst. Tech. J. **48**, 2909 (1969)
33. E.K. Sichel, J.I. Pankove, J. Phys. Chem. Solids **38**, 330 (1977)
34. C.-Y. Luo, H. Marchand, D.R. Clarke, S.P. DenBaars, Appl. Phys. Lett. **75**, 4151 (1999)
35. V. M. Asnin, Fred H. Pollak, J. Ramer, M. Schurman, and I. Ferguson: Appl. Phys. Lett.**75**, 1240 (1999)
36. D. I. Florescu, V. M. Asnin, Fred H. Pollak, R. J. Molnar, and C. E. C. Wood: Appl. Phys. Lett.**88**, 3295 (2000)
37. D. I. Florescu, V. M. Asnin, Fred H. Pollak, A. M. Jones, J. C. Ramer, M. J. Schurman, and I. Ferguson: Appl. Phys. Lett.**77**, 1464 (2000)
38. T. Miyajima, M. Ozawa, T. Asatsuma, H. Kawai, and M. Ikeda: J. Crystal Growth **189/190**, 768 (1998)
39. T. Izumi, Y. Narukawa, K. Okamoto, Y. Kawakami, Sg. Fujita, and S. Nakamura: J. Luminescence **87**, 119 (2000)
40. Ž. Podlipskas, J. Jurkevičius, A. Kadys, S. Miasojedovas, T. Malinauskas, R. Aleksiejūnas, Sci. Rep. **9**, 17346 (2019)
41. Z.T. Ye, H.T. Nguyen, S.-W. Feng, H.-C. Wang, H.-L. Chou, Appl. Sci. **9**, 2279 (2019)
42. P. Ščajev, K. Jarašiūnas, and J. Leach: J. Appl. Phys. **127**, 245705 (2020)
43. K. Okamoto, K. Inoue, Y. Kawakami, S. Fujita, M. Terazima, A. Tsujimura, I. Kidoguchi, Rev. Sci. Instrum. **74**, 575 (2003)
44. H. Murotani, Y. Yamada, Y. Honda, H. Amano, Phys. Status Solidi B **252**, 940 (2015)
45. K. Kojima, T. Ohtomo, K. Ikemura, Y. Yamazaki, M. Saito, H. Ikeda, K. Fujito, S.F. Chichibu, J. Appl. Phys. **120**, 015704 (2016)
46. A.A. Yamaguchi, K. Kawakami, N. Shimizu, Y. Takahashi, G. Kobayashi, T. Nakano, S. Sakai, Y. Kanitani, S. Tomita, IEICE Trans. Electron. **E101**, 527 (2018)
47. Y. Kawakami, K. Inoue, A. Kaneta, K. Okamoto, M. Funato, J. Appl. Phys. **117**, 105702 (2015)
48. K. Okamoto, S. Fujita, Y. Kawakami, A. Scherer, Phys. Stat. Sol. B **240**, 368 (2003)
49. K. Okamoto, A. Scherer, Y. Kawakami, Appl. Phys. Lett. **87**, 161104 (2005)
50. Y. Kawakami, A. Kaneta, L. Su, Y. Zhu, K. Okamoto, M. Funato, A. Kikuchi, K. Kishino, J. Appl. Phys. **107**, 023522 (2010)
51. A. Scherer, J.L. Jewell, Y.H. Lee, J.P. Harbison, L.T. Florez, Appl. Phys. Lett. **55**, 2724 (1989)
52. M. Nirmal, B.O. Dabbousi, M.G. Bawendi, J.J. Macklin, J.K. Trautman, T.D. Harris, L.E. Brus, Nature **383**, 802 (1996)
53. R. Micheletto, M. Abiko, A. Kaneta, Y. Kawakami, Y. Narukawa, T. Mukai, Appl. Phys. Lett. **88**, 061118 (2006)
54. E.M. Purcell, Phy. Rev. **69**, 681 (1946)
55. K. Okamoto, M. Lončar, T. Yoshie, A. Scherer, Y. Qiu, P. Gogna, Appl. Phys. Lett. **82**, 1676 (2003)
56. T. Yoshie, M. Lončar, K. Okamoto, Y. Qiu, O.B. Shchekin, H. Chen, D.G. Deppe, A. Scherer, Proc. SPIE **5360**, 16 (2004)
57. W.L. Barnes, A. Dereux, T.W. Ebbesen, Nature **424**, 824 (2003)
58. S.C. Kitson, W.L. Barnes, J.R. Sambles, Phys. Rev. Lett. **77**, 2670 (1996)
59. J. Takahara, S. Yamagishi, H. Taki, A. Morimoto, T. Kobayashi, Opt. Lett. **22**, 475 (1997)
60. T.W. Ebbesen, H.J. Lezec, H.F. Ghaemi, T. Thio, P.A. Wolff, Nature **391**, 667 (1998)
61. S.A. Maier, M.L. Brongersma, P.G. Kik, S. Meltzer, A.A.G. Requicha, H.A. Atwater, Adv. Mater. **13**, 1501 (2001)
62. H.A. Atwater, Sci. Am. **296**, 56 (2007)
63. I. Gontijo, M. Boroditsky, E. Yablonovitch, S. Keller, U.K. Mishra, S.P. DenBaars, Phy. Rev. B **60**, 11564 (1999)
64. A. Neogi, C.-W. Lee, H.O. Everitt, T. Kuroda, A. Tackeuchi, E. Yablonovitch, Phy. Rev. B **66**, 153305 (2002)

65. K. Okamoto, I. Niki, A. Shvartser, Y. Narukawa, T. Mukai, A. Scherer, Nature Materials **3**, 601 (2004)
66. K. Okamoto, M. Funato, Y. Kawakami, K. Tamada, J. Photochem. Photobiol. C: Photochem. Rev. **32**, 58 (2017)
67. H. Raether, *Surface Plasmons on Smooth and Rough Surfaces and on Gratings* (Springer-Verlag, 1988)
68. K. Okamoto, Y. Kawakami, Phys. Stat. Solidi C **7**, 2582 (2010)
69. K. Okamoto, I. Niki, A. Scherer, Y. Narukawa, T. Mukai, Y. Kawakami, Appl. Phys. Lett. **87**, 071102 (2005)
70. K. Okamoto, J. Jpn. Assoc. Crystal Growth **41**, 162 (2014)
71. X. Xu, M. Funato, Y. Kawakami, K. Okamoto, K. Tamada, Opt. Exp. **21**, 3145 (2013)
72. R.F. Oulton, V.J. Sorger, T. Zentgraf, R.M. Ma, C. Gladden, L. Dai, G. Bartal, X. Zhang, Nature **461**, 629 (2009)
73. Y.J. Lu, J. Kim, H.Y. Chen, C.H. Wu, N. Dabidian, C.E. Sanders, C.Y. Wang, M.Y. Lu, B.H. Li, X.G. Qiu, W.H. Chang, L.J. Chen, G. Shvets, C.K. Shih, S. Gwo, Science **337**, 450 (2012)
74. M.A. Noginov, G. Zhu, A.M. Belgrave, R. Bakker, V.M. Shalaev, E.E. Narimanov, S. Stout, E. Herz, T. Suteewong, U. Wiesner, Nature **460**, 1110 (2009)
75. X. Meng, A.V. Kildishev, K. Fujita, K. Tanaka, V.M. Shalaev, Nano Lett **13**, 4106 (2013)
76. K. Okamoto, K. Okura, P. Wang, S. Ryuzaki, K. Tamada, Nanophotonics **9**, 3409 (2020)
77. X.F. Gu, T. Qiu, W.J. Zhang, P.K. Chu, Nanoscale Res. Lett. **6**, 199 (2011)
78. G. Lozano, S.R.K. Rodriguez, M.A. Verschuuren, J. Gomez Rivas: Light: Science & Applications **5**, e16080 (2016)
79. N. Okada, T. Tsukada, K. Tateishi, K. Okamoto, K. Tadatomo, J. Appl. Phys. **121**, 153102 (2017)
80. C.-Y. Chen, Y.-C. Lu, D.-M. Yeh, C.C. Yang, Appl. Phys. Lett. **90**, 183114 (2007)
81. D.-M. Yeh, C.-F. Huang, Y.-C. Lu, C.-Y. Chen, T.-Y. Tang, J.-J. Huang, K.-C. Shen, Y.-J. Yang, C.C. Yang, Appl. Phys. Lett. **91**, 063121 (2007)
82. C.-F. Lu, C.-H. Liao, C.-Y. Chen, C. Hsieh, Y.-W. Kiang, C.C. Yang, Appl. Phys. Lett. **96**, 261104 (2010)
83. W. Yang, Y. He, L. Liu, X. Hu, Appl. Phys. Lett. **102**, 241111 (2013)
84. Takada, T. Ohto, R.G. Banal, K. Okamoto, M. Funato, Y. Kawakami: 57th JSAP Spring Meeting, 17p-TC-13 (2010)
85. J. Lin, A. Mohammadizia, A. Neogi, H. Morkoc, M. Ohtsu, Appl. Phys. Lett. **97**, 221104 (2010)
86. C.-Y. Cho, Y. Zhang, E. Cicek, B. Rahnema, Y. Bai, R. McClintock, M. Razeghi, Appl. Phys. Lett. **102**, 211110 (2013)
87. N. Gao, K. Huang, J.C. Li, S.P. Li, X. Yang, J.Y. Kang, Sci. Rep. **2**, 816 (2012)
88. K. Okamoto, K. Tateishi, K. Tamada, M. Funato, and Y. Kawakami: Jpn. J. Appl. Phys.**58**, SCCB31 (2019)
89. K. Tateishi, P.P. Wang, S. Ryuzaki, M. Funato, Y. Kawakami, K. Okamoto, K. Tamada, Appl Phys Lett. **111**, 172105 (2017)
90. F.D. Sala, A.D. Carlo, P. Lugli, F. Bernardini, V. Fiorentini, R. Scholz, J. Jancu, Appl Phys Lett. **74**, 2002 (1999)
91. T. Kuroda, A. Tackeuchi, T. Sota, Appl Phys Lett. **76**, 3753 (2000)
92. K. Tateishi, M. Funato, Y. Kawakami, K. Okamoto, K. Tamada, Appl Phys Lett. **106**, 121112 (2015)
93. K. Tateishi, P. Wang, S. Ryuzaki, M. Funato, Y. Kawakami, K. Okamoto, K. Tamada, Appl. Phys. Exp. **11**, 052016 (2019)

Chapter 2
Ultrafast Infrared Plasmonics—A Novel Platform for Spectroscopy and Quantum Control

Satoshi Ashihara and Ikki Morichika

Abstract A marriage of ultrashort-pulsed laser and plasmonics in the mid-infrared range creates a novel platform for light–molecule interactions. Resonant gold nanoantennas are exploited to realize ultrasensitive nonlinear vibrational spectroscopy and vibrationally mediated chemical reaction control. Arbitrary shaping of spatiotemporal field distribution both in nanometer scale and in femtosecond scale has a great promise for advanced vibrational spectroscopy and controlled ground-state chemistry.

2.1 Ultrafast Plasmonics in the Mid-Infrared

2.1.1 Nanoscale Localization of Ultrashort Pulses

An ultrashort laser pulse is characterized by its short temporal duration, wide spectral bandwidth, and high peak power. These features are particularly beneficial for the application of nonlinear spectroscopy, which enables time-resolved measurements of ultrafast processes in atoms, molecules, and solids.

Since the time scale of nuclear motion in molecules and solids is close to or longer than 10 fs (10^{-14} s), ultrashort laser pulses with femtosecond time duration have provided insight into nuclear motions in chemical reactions, opening an innovative field called 'femtochemistry' [1].

Ultrashort laser pulses are useful not only for observing ultrafast processes but also for actively controlling them. Many spectral components with well-defined phase

S. Ashihara (✉) · I. Morichika
Institute of Industrial Science, The University of Tokyo, Meguro-ku, Tokyo, Japan
e-mail: ashihara@iis.u-tokyo.ac.jp

I. Morichika
e-mail: m-ikki@iis.u-tokyo.ac.jp

T. Yatsui (ed.), *Progress in Nanophotonics 6*, Nano-Optics and Nanophotonics,
https://doi.org/10.1007/978-3-030-71516-8_2

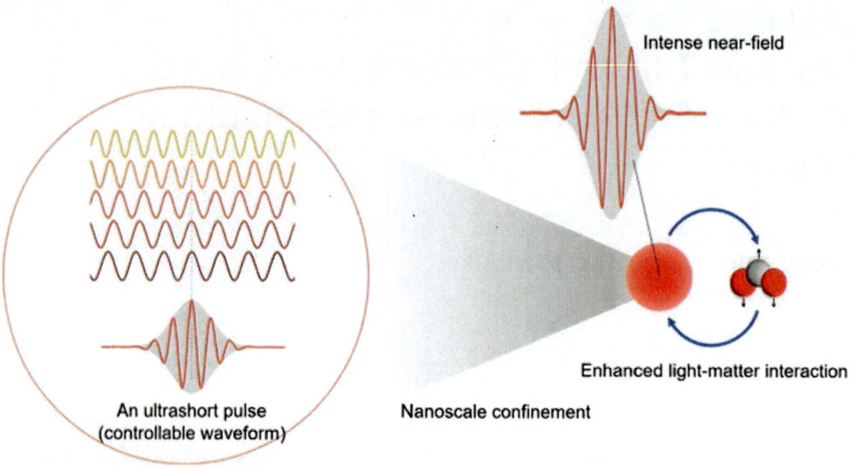

Fig. 2.1 Concept of nanoscale localization of an ultrashort optical pulse

relationships that constitute ultrashort pulses enable coherent excitation of multiple electronic, vibrational, and rotational transitions and manipulation of quantum mechanical interference of transition pathways [2, 3]. Such control over the microscopic behavior of matter, called quantum control, is essential for chemical reaction control and quantum information processing [4–6].

When an ultrashort optical pulse illuminates a metal nanostructure, a surface plasmon is excited, and an intense electromagnetic field, localized both in space and in time, is generated [7] (see Fig. 2.1). Most importantly, the degrees of freedom in the temporal waveform of an ultrashort pulse and the degrees of freedom in the geometrical shape of the nanostructure allow us *to control an electric-field distribution both in nanometer scale and in femtosecond scale* [7–12]. Such spatiotemporally confined and designable fields have great promise for substantial enhancement and precise control of light–matter interactions.

2.1.2 Light–Matter Interaction in the Mid-Infrared Range

An electromagnetic wave in the mid-infrared (mid-IR) range (2 to 20 μm) resonates with a variety of molecular vibrational modes, making it a powerful probe for structural analysis and sensing of chemical species. The most popular type of linear infrared (IR) vibrational spectroscopy is Fourier transform infrared (FT-IR) spectroscopy using a thermal light source.

By utilizing ultrashort mid-IR pulses to exploit nonlinear vibrational spectroscopy, such as pump–probe spectroscopy and two-dimensional spectroscopy, one can resolve ultrafast dynamics of structural changes and vibrational relaxations, as well as vibrational anharmonic coupling, which cannot be revealed by linear IR spectroscopy [13, 14]. In fact, the structure and dynamics of hydrogen-bonded molecular systems which are deeply involved in life (fluctuating structure of liquid water, folding dynamics of proteins, etc.) have been revealed by nonlinear vibrational spectroscopy [15, 16].

Ultrashort mid-IR pulses enable mode-selective, multi-quantum vibrational excitation with broad accessibility to the potential energy surface. They provide a powerful tool in active control over molecular processes at electronic ground states because the strong excitation of key reagent vibrations directly contributes to selective reaction control [17], as has been validated by theoretical studies [18–21].

However, nonlinear vibrational spectroscopy and vibrational quantum control have suffered from low sensitivity and low efficiency, respectively, mainly due to the small dipole moments of vibrational transitions (they are generally much smaller than those of electronic transitions). In fact, the observation of nonlinear vibrational signals from a small number of molecules or for vibrational modes of smaller transition dipole moments is challenging. In chemical reaction control experiments, multi-quantum vibrational excitation is usually inefficient, especially in condensed phases, and therefore, energy injected into a specific bond does not exceed the potential barrier of the targeted reaction.

2.1.3 A Novel Platform for Vibrational Spectroscopy and Quantum Control

We may resolve these issues via a marriage of ultrafast laser technology and plasmonics in the mid-IR range. Surface plasmon excitation enables subwavelength localization and strong field enhancement of ultrashort mid-IR pulses, which is expected to strengthen the interactions of mid-IR pulses with molecular vibrations, creating a novel platform for nonlinear vibrational spectroscopy and vibrational quantum control.

In recent years, we have studied ultrafast responses and spatial characteristics of surface plasmons in the mid-IR range [22–24] and performed non-destructive measurements of plasmonic near-fields on a femtosecond timescale by using a nonlinear photoemission process at a metal surface [25]. Then, we applied the mid-IR plasmonics to realize enhanced strong-field phenomena at the nanoscale [25, 26], ultrasensitive nonlinear vibrational spectroscopy [27, 28], and vibration-mediated chemical reaction control [29].

2.1.4 *Outline*

The outline of this chapter is as follows. We describe the basic properties of surface plasmons and metal nanoantennas in the mid-IR range in Sect. 2.2, we describe the measurement of plasmonic near-fields by means of field-driven photoemission in Sect. 2.3, and we present the applications of resonant metal nanoantennas to ultrafast vibrational spectroscopy and chemical reaction control in Sects. 2.4 and 2.5, respectively. We summarize and provide future perspectives in Sect. 2.6.

2.2 Mid-IR Plasmonics

2.2.1 *Surface Plasmon Polariton on Gold*

Plasmonics in the mid-IR range is gaining increasing attention [30–32] because of promising applications in surface-enhanced spectroscopy [33–39], chemical/biosensing [40, 41], thermal radiation control [42], optoelectronic circuits [43, 44], etc. Surface plasmons may be excited at mid-IR wavelengths on various materials [31, 32], but their properties are different according to the material in which they are excited. In this subsection, we briefly review surface plasmons on gold at mid-IR wavelengths by comparing them with those on a doped semiconductor.

When a slowly varying electric field is applied to a conducting material, free electrons move and thus induce electric polarization in the direction opposite to the external field. This motion is reflected in a large negative real part of the permittivity. When the electric field oscillates at a higher frequency, free electrons do not immediately follow the change in the electric field. As a result, the absolute value of the permittivity decreases. This motion of free electrons and the resultant change in permittivity govern the frequency-dependent characteristics of surface plasmons.

Figure 2.2 shows the dispersion relationship of surface plasmon polaritons (SPPs) propagating along the interface between gold and air (red line) at a frequency below 100 THz or a wavelength above 3 μm (the Drude parameters are taken from [45]). As seen in the inset of Fig. 2.2, the dispersion relation bends strongly in the visible range as the frequency approaches the plasma frequency of gold. The decreased slope of the dispersion curve indicates that the phase velocity and wavelength decrease (that is, wavelength compression occurs). In contrast, the dispersion curve is close to the light line (gray dashed line) in the near- and mid-IR ranges, which indicates that the phase velocity is close to that of free-space light in air.

Figure 2.3a shows the calculated penetration depth of the electromagnetic field of an SPP into air, δ_d, normalized by the free-space wavelength λ_0. The electromagnetic field of the SPP is localized within the subwavelength dimension from the interface in the visible range, whereas it penetrates into air with a dimension much larger than λ_0 in the mid-IR range. Figure 2.3b shows the calculated propagation length

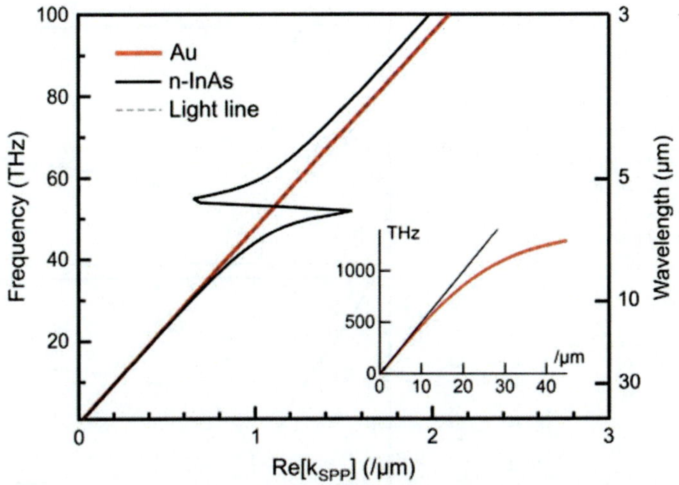

Fig. 2.2 Dispersion curve of SPP propagating along Au/air interface [45] (*red line*) and that of SPP propagating along *n*-InAs/air interface [46] (*black line*). A light line is shown as *gray dashed line*

L_{SPP} normalized by λ_0. We can see that the propagation length increases with the operation wavelength, exceeding 10^3 times the free-space wavelength λ_0.

As mentioned above, SPPs behave differently depending on the operation wavelength. In the visible range, the electromagnetic field is strongly bound to the interface, the wavelength compression is remarkable, and the propagation distance is short. In the near- and mid-IR ranges, the electromagnetic field is only weakly bound to the interface, the wavelength compression is subtle, and the propagation distance is long.

To obtain subwavelength mode confinement in the mid-IR range, materials with plasma frequencies in the mid-IR range should be exploited. As an example, we display in Fig. 2.2a a dispersion curve of *n*-doped InAs with a plasma frequency of 5.5 μm [46]. The dispersion curve bends at mid-IR wavelengths of 6–10 μm, indicating wavelength compression at these wavelengths. The penetration depth δ_{SPP} is smaller than the free-space wavelength λ_0, as shown in Fig. 2.3a, meaning that subwavelength confinement can be achieved in the mid-IR range. However, there is a cost of substantially reduced propagation length L_{SPP} as shown in Fig. 2.3b.

A good reference to the achievable electric-field enhancement is a figure of merit (FOM) defined as $M = L_{SPP}/\delta_{SPP}$, which provides a clear illustration of the confinement/propagation length trade-off [32]. As we see from Fig. 2.3a, b, gold exhibits a much larger FOM than *n*-InAs at mid-IR wavelengths. As this comparison shows, noble metals are generally more suited for applications in which a small loss, long propagation length, or large electric-field enhancement is required in the mid-IR range than materials that have plasma frequencies in a lower frequency range. Although the noble metals of silver and copper display very similar SPP properties, gold is an excellent plasmonic material because of its superior chemical stability [47].

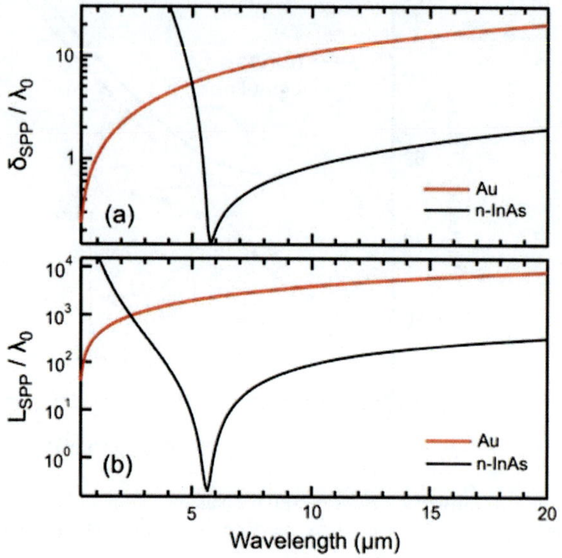

Fig. 2.3 **a** Penetration depth of the SPP into air (normalized to the free-space wavelength λ_0) at Au/air interface (*red*) and at *n*-InAs/air interface (*black*). **b** Propagation length (normalized to λ_0) of SPP at Au/air interface (*red*) and at *n*-InAs/air interface (*black*)

2.2.2 Gold Nanoantennas

Metal nanorods serve as dipole antennas in the visible and IR ranges. They may be viewed as plasmonic cavities where SPPs form stationary or eigenmodes. Therefore, the resonance frequency may be tailored by adjusting the size, shape, and dielectric properties of the nanorod. When a nanorod is resonantly excited, the electromagnetic field is significantly enhanced near the nanorod ends. This electric-field enhancement is useful for strengthening light–matter interactions, especially nonlinear processes that increase nonlinearly with the field strength.

In applying plasmonic near-field enhancements to nonlinear spectroscopy and quantum control where ultrashort pulses are exploited, not only the magnitude of electric-field enhancement but also its temporal/spectral response is important. In this subsection, we describe the spectral properties of near-field enhancement obtainable with gold nanorods, which resonate at mid-IR frequencies [22].

Figure 2.4 displays a typical electric-field distribution or field enhancement factor for a resonantly excited gold nanorod (1250 nm in length) on a ZnS substrate, as calculated by using the finite difference time-domain (FDTD) method. The electric field is substantially enhanced near the nanorod ends in a dimension much smaller than the free-space wavelength.

Electric-field enhancement at a frequency ν is expressed with a transfer function defined as $H(\nu) = E_{\mathrm{lo}}(\nu)/E_{\mathrm{in}}(\nu)$, where $E_{\mathrm{in}}(\nu)$ and $E_{\mathrm{lo}}(\nu)$ denote the complex electric-field amplitude of the incident light and the enhanced near-field, respectively. Correspondingly, the intensity enhancement factor and the enhancement phase are defined as $\eta(\nu) = |H(\nu)|^2$ and $\phi(\nu) = \arg H(\nu)$, respectively.

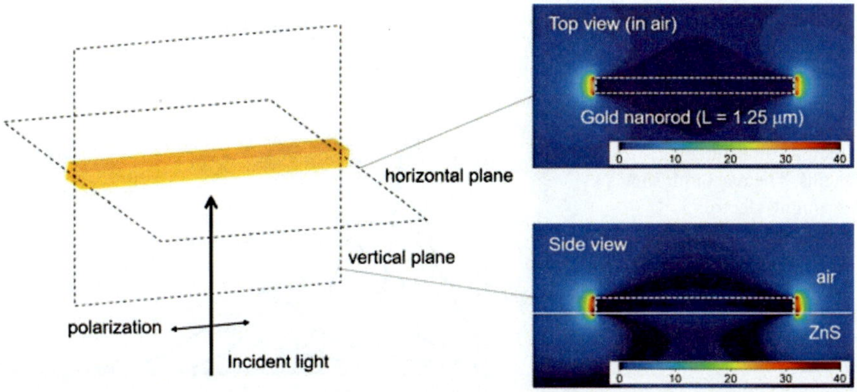

Fig. 2.4 Simulated near-field distribution at each of the horizontal and vertical cross sections when a gold nanorod with a length of 1.25 μm on a ZnS substrate is resonantly excited

The intensity enhancement factor $\eta(\nu)$ at a position near the nanorod tip is displayed in Fig. 2.5a for gold nanorods with varied lengths ($L = 1000$, 1250, 1750, and 2500 nm) and a common width (thickness) of 100 nm (100 nm) [22]. Here, we estimate $H(\nu)$ for the electric-field component parallel to the nanorod long axis. Each nanorod exhibits near-field enhancement at its resonant frequency, and the enhancement factor at resonance ranges from 300 to 1500.

Figure 2.5b shows the extinction cross section of the gold nanorod obtained from the transmission measurement performed with an FT-IR spectrometer [22]. Here, gold nanorods with the same dimensions as those used in simulations are fabricated on a ZnS substrate. The transmission is measured for the polarization parallel to the nanorod long axis using a polarizer. A typical scanning electron micrograph of a nanorod is shown in the inset. Each nanorod exhibits resonant behavior in the extinction in the same fashion as in the field enhancement. The slight difference in the peak position and the spectral profile between the field enhancement (near-field spectrum) and the extinction (far-field spectrum) is attributed to a general feature of damped harmonic oscillators [48].

Interestingly, Fig. 2.5a, b show that the linewidth $\Delta\nu$ is linearly proportional to the resonance frequency ν_0 of the antenna. This means that the quality factor $\Delta\nu/\nu_0$ of the plasmonic cavity is nearly constant for the nanorods resonating at mid-IR frequencies. This frequency dependence of the linewidth, $\Delta\nu \propto \nu_0$, is explained as the nature of radiation loss (see [22] for details). Since the electromagnetic field of a surface plasmon on gold is only weakly bound to the interface in the mid-IR range, Joule heating loss is suppressed so that damping of the antenna mode is dominated by radiation loss.

For an incidence of an ultrashort pulse, the spectrum of the enhanced near-field may be squeezed and its temporal waveform may be stretched owing to the finite bandwidth of the near-field enhancement. Figure 2.6a shows the enhancement factor $\eta(\nu)$ and the phase $\phi(\nu)$ for a gold nanorod with a length of 1000 nm. The maximum

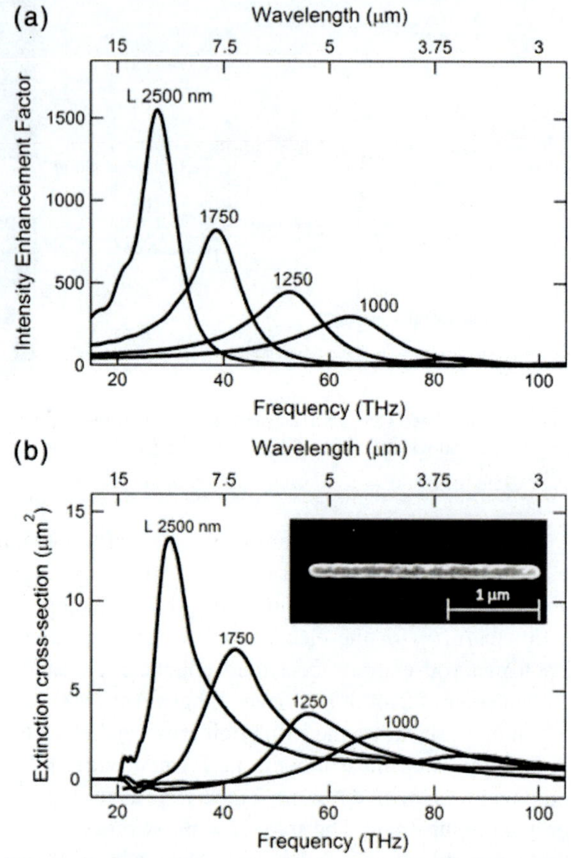

Fig. 2.5 **a** Calculated intensity enhancement factors near the nanorod end and **b** the measured extinction cross sections for nanorods of different lengths. The inset in **b** shows a scanning electron micrograph of a fabricated nanorod. These figures are reprinted with permission from [22]. Copyright ©2014, American Institute of Physics

enhancement is reached at 5 μm and the phase changes from 0 to π across the resonance. For reference, the power spectrum of a 100-fs pulse (dashed line) and that of a 30-fs pulse (dotted line) are also shown in Fig. 2.6a. The frequency spread of the enhancement $\eta(\nu)$ is sufficiently larger than the spectral width of the 100-fs pulse, and the phase ϕ can be approximated as a straight line throughout its pulse spectrum. In contrast, the frequency spread of $\eta(\nu)$ is approximately the same as the spectral width of the 30-fs pulse, and the phase ϕ can no longer be regarded as a straight line.

Figure 2.6b displays the temporal waveform of the enhanced near-field for incident pulses with a duration of 100 fs (upper panel) and 30 fs (lower panel). Here, the incident waveform is shown as a dashed line. For the 100-fs incident pulse, the enhanced near-field faithfully follows the incident waveform except for a phase shift of approximately $\pi/2$. For the 30-fs incident pulse (corresponding to two optical cycles), the enhanced near-field is elongated in time with respect to the incident field. From a series of simulations, we find that near-field enhancement is achieved

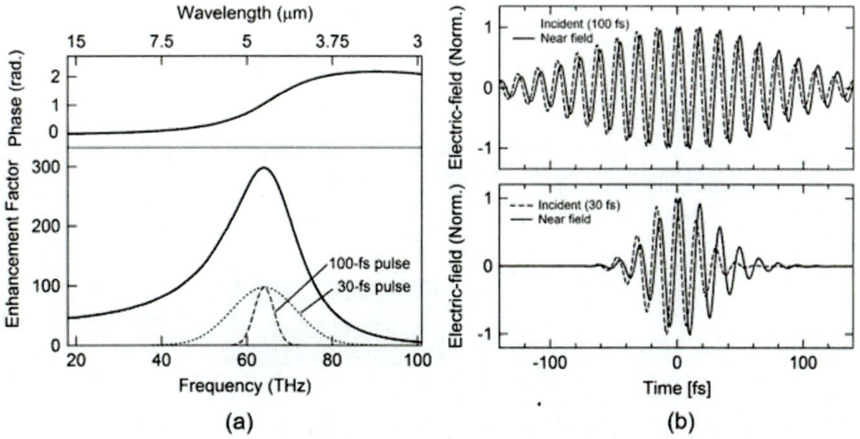

Fig. 2.6 **a** Calculated intensity enhancement factor (*lower panel*) and phase (*upper panel*) for a 1000-nm-long gold nanorod. Power spectra for a 100-fs pulse and a 30-fs pulse are shown as *dashed* and *dotted lines*, respectively. **b** Temporal waveform of the incident electric field (*dashed*) and the enhanced near-field (*solid*) for an incident pulse duration of 100 fs (*upper panel*) and 30 fs (*lower panel*). These figures are reprinted with permission from [22]. Copyright ©2014, American Institute of Physics

with high fidelity in the temporal waveform if the incident pulse contains at least three optical cycles in its envelope [22].

2.3 Measurement of Plasmonic Near-Field Using Field-Driven Photoemission

In this section, we discuss the emission of an electron from gold nanorods driven by mid-IR pulses [25]. Upon resonant excitation of gold nanorods, the enhanced near-field created near the nanorod ends modifies the potential so that free electrons in gold may be emitted via quantum mechanical tunneling (see Fig. 2.7). Because the tunneling current exhibits characteristic dependence on the local electric field, it is possible to extract local field strength from intensity-dependent tunneling current measurements. Such quantitative information on the plasmonic near-field is of immediate relevance for antenna-enhanced nonlinear vibrational spectroscopy and antenna-enhanced quantum control, which will be described in the following sections.

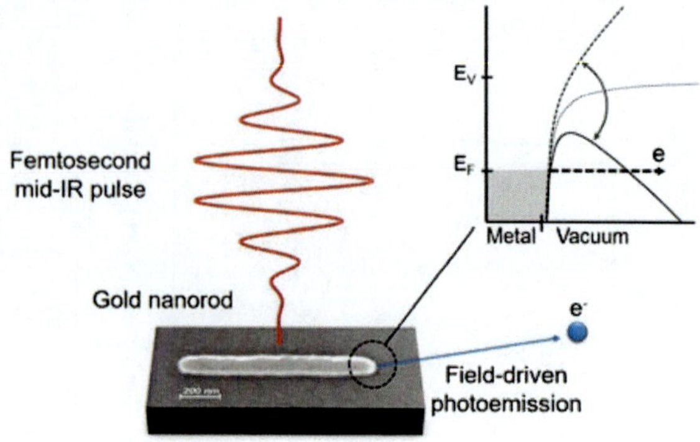

Fig. 2.7 Schematic of field-driven photoemission from a gold nanorod. A free electron in a gold nanorod may be emitted via quantum mechanical tunneling because of a modification of a potential induced by a plasmonic near-field of a mid-IR pulse

2.3.1 Field-Driven Photoemission

Strong-field light–matter interaction at surfaces and nanostructures has been intensively studied to date, with special focus on local field enhancements and subwavelength confinement of optical fields. Nonlinear photoelectron emission has been studied for various metallic nanostructures [49–59], with motivations for ultrafast electron [60–62] and X-ray sources [63], phase-resolved imaging and spectroscopy [64, 65], etc.

Metal nanotips, displaying broadband near-field enhancements, exhibit a representative model system for nonlinear photoelectron emission and acceleration [49, 51, 54, 64, 66]. Nanoparticles and plasmonic antennas, exhibiting field enhancements from surface plasmon resonance and geometric edge enhancement, constitute another platform for photoemission studies [55–57, 59].

When a metal is illuminated by light with a photon energy larger than the work function of the metal, conducting electrons inside the metal can be emitted (photoelectric effect). Even if the photon energy is smaller than the work function, electrons may be emitted via multiphoton absorption or via quantum mechanical tunneling if the illuminating light is intense enough [67]. In the former mechanism, the electron emission rate follows the power-law scaling reflecting the multiphoton process. In the tunneling mechanism, the emission rate follows the physics of quantum mechanical tunneling as we describe later, and interestingly, electron emission occurs within a half cycle of a driving optical pulse [67].

Nonlinear photoemission is well described by the multiphoton mechanism at a lower intensity of $\gamma > 1$ and by the tunneling mechanism at a higher intensity of $\gamma < 1$ [68]. Here, γ is a Keldysh parameter defined as

$$\gamma = \sqrt{\frac{\phi}{2U_p}}, \tag{2.1}$$

where ϕ is a metal work function and $U_p = e^2F^2/(4m\omega^2)$ is ponderomotive energy (e and m are the electron charge and mass, respectively). As we can see, the Keldysh parameter γ is proportional to the optical frequency ω and inversely proportional to the electric field amplitude F. Here, we see that mid-IR pulses provide higher ponderomotive energies and therefore enable access to the strong-field regime at reduced local intensities [54, 58], compared with visible pulses.

2.3.2 Nonlinear Photoemission Experiments

A gold nanorod 1 μm in length, 150 nm in width, and 50 nm in height is patterned into two-dimensional arrays on a ZnS substrate, as shown in Fig. 2.8a, by using electron beam lithography and a lift-off process. Here, the near-field coupling between neighboring nanorods is prevented by a spacing of 5 μm. A 5-nm-thick gold capping serves as a conduction layer to avoid sample charge-up during photoemission.

Mid-IR pulses at center wavelengths tunable from 3–10 μm and pulse durations of approximately 160 fs are produced at a 1 kHz repetition rate by parametric down conversion of amplified Ti:sapphire laser pulses. A parabolic mirror is used to focus the mid-IR pulses to spot diameters of approximately 30 μm. To illuminate multiple rods (approximately 8000), the sample is placed at a defocus position with a beam diameter of approximately 500 μm, giving rise to incident peak intensities of 0.5–1.5 GW/cm^2 for pulse energies of 80–230 nJ. The excitation linear polarization is set to be parallel to the nanorod axis, and we excite a fundamental longitudinal mode or the half-wave dipole antenna mode that has a resonance wavelength of 5 μm. The emitted photoelectrons are guided by a magnetic bottle to a time-of-flight (TOF) spectrometer for measuring photoelectron yields and kinetic energy spectra.

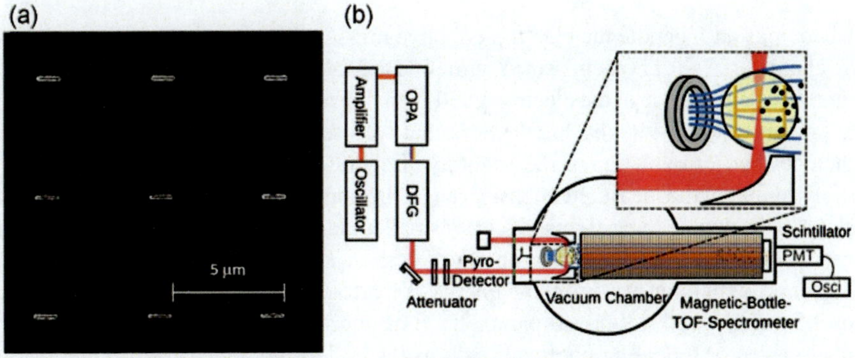

Fig. 2.8 **a** Scanning electron micrograph of a nanorod array (rod length 1 μm). **b** Experimental setup. Tunable mid-IR pulses are focused onto a nanorod array in a high vacuum chamber. Electron kinetic energy spectra are measured with a magnetic-bottle time-of-flight spectrometer. These figures are reprinted with permission from [25]. Copyright ©2015, American Institute of Physics

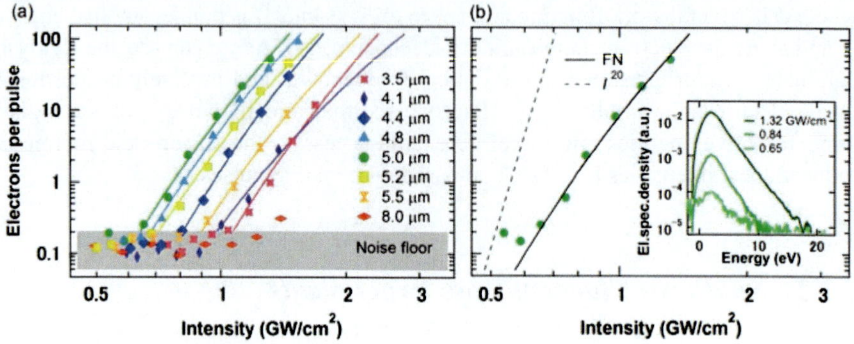

Fig. 2.9 **a** Electron yield for different excitation wavelengths as a function of incident intensity. Each curve is fitted by Fowler–Nordheim equation (*solid line*) to extract the intensity enhancement factor. **b** Fowler–Nordheim fit to the electron yield at 5 µm wavelength (*solid line*). For comparison, a corresponding multiphoton dependence (*dashed line*, slope 20) is shown. Inset: Electron kinetic energy spectra for selected intensities. These figures are reprinted with permission from [25]. Copyright ©2015, American Institute of Physics

The photoelectron yield is displayed as a function of incident peak intensity for varied excitation wavelengths in Fig. 2.9a. All curves exhibit a strongly nonlinear intensity dependence. We also find that the curves for excitation wavelengths closer to the antenna resonance reside at lower intensity, which indicates *the underlying near-field enhancement*.

Under the static-field approximation, the tunnel current density J generated by an electric field F at the metal surface is given by the Fowler–Nordheim (FN) equation [69]

$$J = \frac{e^3 E^2}{16\pi^2 \hbar \phi} \exp\left(-\frac{4\sqrt{2m_e}\phi^{3/2}}{3\hbar F}\right), \tag{2.2}$$

where m_e and $\hbar$ denote the electron effective mass and Dirac constant, respectively. To characterize the photoemission regime and to evaluate local field enhancement, the intensity dependence of the electron yield is analyzed by fitting with the FN equation, as shown in Fig. 2.9 (solid lines). Note that we use the modified FN equation that includes barrier reduction or the Schottky effect [70].

Throughout the measured intensity range, the nonlinearity of the yield (or the local slope in the double logarithmistic plot) is smaller than the corresponding multiphoton dependence (slope 20 at a photon energy of 0.25 eV and the 5.1 eV work function) and decreases with intensity. From the FN fits, we extract a local electric-field amplitude of 3.5 Vnm^{-1} and a Keldysh parameter γ of approximately 0.80, which indicates field-driven or tunneling photoemission as the underlying mechanism. The kinetic energy spectra of photoelectrons measured for varied incident intensities at 5 µm excitation are displayed in inset of Fig. 2.9b. We find that the kinetic energy cutoff (when defined to include 99% of the electron population) linearly increases with intensity, consistent with ponderomotive scaling [71].

2.3.3 Estimation of Plasmonic Near-Field

Fig. 2.10a displays the measured far-field extinction spectrum of the nanorod arrays, which exhibit a resonance at a wavelength of 5 μm. Figure 2.10b displays the photoelectron yield measured for different excitation wavelengths at a constant incident intensity of 1.1 GW/cm^2 (triangles). The photoelectron yield peaks at 5 μm, indicating that the observed photoemission is governed by the resonantly enhanced near-field. The intensity enhancement factor obtained from the FN fits to the electron yields is plotted as circles in Fig. 2.10b, where we find a maximum intensity enhancement factor of 1290 and the corresponding maximum field enhancement factor of 36. The photoelectron yield exhibits much sharper resonance profile compared with the underlying near-field, which is reasonably understood as a result of the high nonlinearity of the emission process.

To support the experimental results, the electromagnetic field distribution is numerically calculated with the FDTD method [22]. Here, the nanorods are modeled with the experimental dimensions (length: 1 μm, height: 50 nm, width: 150 nm), as shown in the inset of Fig. 2.10c. The simulated far-field extinction shown in Fig. 2.10c agrees well with the experiment, exhibiting an asymmetric resonance at 5 μm with a linewidth of 21 THz. The inset in Fig. 2.10c shows the spatial distribution of the electric-field component parallel to the nanorod axis for 5 μm excitation. The inten-

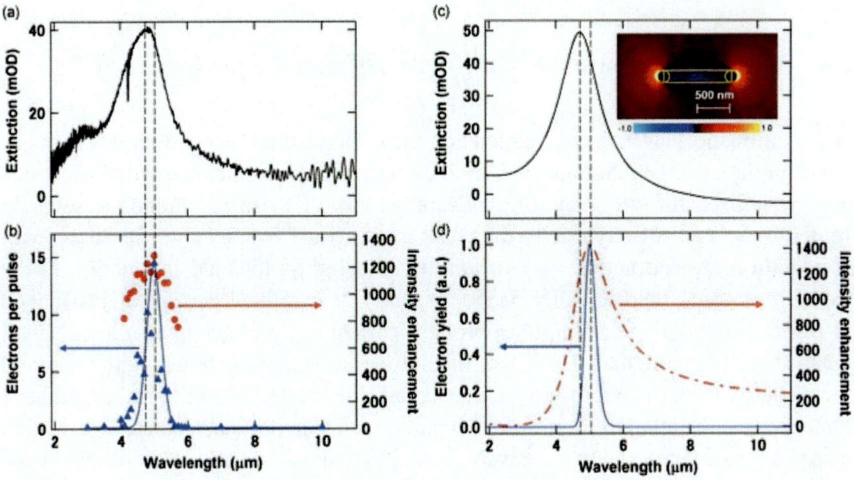

Fig. 2.10 **a** Measured far-field extinction spectrum. **b** Experimental photoemission yield as a function of excitation wavelength and for constant incident intensity (*triangles*) and intensity enhancement factors extracted from the corresponding yield curves (*circles*). **c** Simulated far-field extinction spectrum for 1 μm long nanorods. The inset shows the electric-field distribution for the component parallel to the nanorod. **d** Simulated local intensity enhancement factor (*dashed-dotted line*) and electron yield (*solid line*) as a function of the excitation wavelength. Dashed vertical lines indicate a redshift of spectral near-field enhancement relative to the far-field extinction spectrum. These figures are reprinted with permission from [25]. Copyright ©2015, American Institute of Physics

sity enhancement factor at the position of the largest electric field is shown as a dashed-dotted line in Fig. 2.10d. Here, we find that the maximum value of approximately 1400 agrees well with the experimental value. The solid line represents the photoelectron yield calculated with the FN equation. It exhibits a narrowed width of 5 THz, in agreement with the experimental value of approximately 6 THz.

In both the experiment and the simulations, the near-field enhancement is highest at a slightly red-shifted wavelength compared to the far-field extinction spectrum (see vertical dashed lines in Fig. 2.10a–d. In the experiments presented above, such spectral shifts between near- and far-field spectra, universal to damped harmonic oscillators [48, 72], are successfully observed by using the nonlinear photoemission process as a noninvasive probe.

2.3.4 *Summary on Field-Driven Photoemission*

Field-driven photoemission is successfully demonstrated with resonant gold nanorods excited by femtosecond mid-IR pulses. The wavelength-dependent, nonlinear photoelectron yield follows the resonance profile of the nanorods, and intensity-dependent measurements allow for a noninvasive determination of the local intensity (field) enhancement factor well exceeding 1000 (30).

2.4 Antenna-Enhanced Nonlinear Infrared Spectroscopy

Mid-IR ultrashort lasers have enabled nonlinear vibrational spectroscopy, which is a powerful, label-free technique providing direct insight into the molecular structures and dynamics. Unfortunately, the widespread use of these techniques is severely limited by low sensitivity due to the weak interaction of molecular vibrations with IR radiation. An attractive way to amplify the light–molecule interaction is the use of plasmonic field enhancements. Recently, we utilized periodically arranged IR-resonant nanoantennas in pump-probe spectroscopy and achieved a local signal enhancement of more than 6 orders of magnitude, substantially reducing the required pump energy from the micro-joule to the nano-joule level [27]. Furthermore, we have demonstrated the reflection type of antenna-enhanced nonlinear spectroscopy, which is useful for surface-selective measurements [28]. In this section, we will review the historical perspective of such plasmon-enhanced IR spectroscopy and then present our experimental results and considerations.

2.4.1 Vibrational Spectroscopy with Infrared Ultrashort Lasers

The mid-IR range of the spectrum corresponds to energies associated with molecular vibrations (Fig. 2.11). Since the vibrational properties are directly linked to the molecular constituents, their chemical bonds, and their configuration, the mid-IR absorption spectra, often referred to as molecular fingerprints, provide a vast amount of structural information about molecules.

In the last quarter century, the advent of mid-IR ultrashort laser techniques has enabled nonlinear vibrational spectroscopy, such as IR pump-probe spectroscopy and two-dimensional infrared (2D-IR) spectroscopy, which provides richer information than is available from linear spectroscopy [13, 14]. In IR pump-probe spectroscopy, ultrashort pulse sequences are used to excite and detect molecular vibrations, enabling time-resolved study of the molecular dynamics on the timescales on which they actually occur (typically femtosecond to picosecond timescales for elementary steps in chemical reactions) [1]. In 2D-IR spectroscopy, mid-IR absorption spectra are spread into two frequency dimensions, providing information on vibrational couplings, spectral diffusion, and chemical exchange.

Despite their fascinating potential, however, the applications of nonlinear vibrational spectroscopy are somewhat limited due to the small absorption cross sections of molecular vibrations ($\sigma_{abs} \approx 10^{-20}\mathrm{cm}^2$). The experiments require mid-IR pump pulses of the micro-joule level, typically generated by the complex frequency conversion system based on the regenerative amplifier. Even if the mid-IR pulses of the micro-joule level are exploited, nonlinear signals can be observed for only a limited number of vibrational modes whose absorption cross sections are relatively large (e.g., CO, NH, OH stretching vibrations). These facts severely limit the widespread use of the nonlinear vibrational spectroscopy.

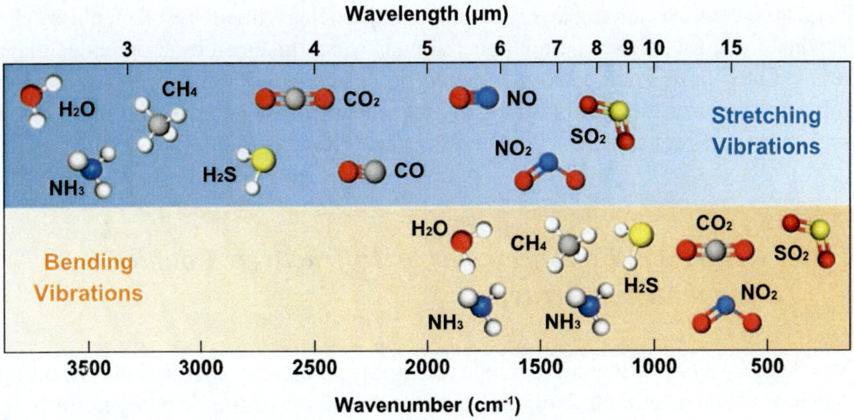

Fig. 2.11 Characteristic molecular vibrations in the mid-infrared range

2.4.2 Surface-Enhanced Infrared Absorption Spectroscopy

One promising approach to overcome the limitations described in the previous subsection is to amplify the interaction of molecular vibrations with IR light utilizing surface plasmon excitation. As described in Sect. 2.2, plasmonics take advantage of the coupling of light to metal electrons, enabling subwavelength localization and strong field enhancements. Therefore, surface plasmon excitation substantially enhances the light–matter interactions.

For linear IR spectroscopy, this concept was first applied in 1980 by Hartstein and coworkers [33]. They demonstrated that linear vibrational signals of molecules adsorbed on randomly arranged Ag nanoparticles (with plasmonic resonances located in the visible spectral range) were enhanced by a factor of 20. This phenomenon, which was later named surface-enhanced infrared absorption (SEIRA), is currently commonly used in various applications such as electrochemistry and biosensing [39, 41].

A significantly stronger enhancement can be obtained by using well-defined metallic nanostructures whose plasmonic resonances are tuned to the respective molecular vibration. This approach of resonant SEIRA was first demonstrated in 2008 by Neubrech and coworkers [34]. They achieved the detection of 50 attograms of molecules located in the plasmonic hot-spots of a nanoantenna and found that the vibrational signals are enhanced by 5 orders of magnitude. In the following years, along with significant advances in nanofabrication technologies, numerous SEIRA morphologies have been proposed and demonstrated to substantially enhance the vibrational signals for an expanding range of applications[37, 38].

Plasmonic near-field enhancement is expected to be useful for amplifying nonlinear vibrational signals as well. In fact, the benefits of near-field enhancement are expected to be larger for nonlinear spectroscopy because nonlinear signals increase superlinearly with field strength. In 2013, Donaldson et al. first applied near-field enhancement to nonlinear IR spectroscopy [73]. They demonstrated 2D-IR spectroscopy on capping layers of gold nanoparticles with off-resonant plasmonic responses and found that the nonlinear signals were enhanced by 2 orders of magnitude. Later, Selig et al. performed pump-probe and 2D-IR spectroscopies of 5 nm polymer layers on randomly arranged IR-resonant nanoantennas and observed a nonlinear signal enhancement of 4 orders of magnitude [74].

2.4.3 Near-Field Enhancements by Diffractively Coupled Nanoantenna Arrays

An obvious way to further increase the sensitivity is to use nanostructures providing higher field enhancements. Near-field enhancement can be improved by optimizing not only the geometries of the individual nanoantennas but also their arrangements. Nanoantenna arrays arranged in a certain period scatter incident light to produce diffracted waves. When one of the diffracted waves propagates in the plane of the array, it may couple the plasmonic resonances associated with individual nanoantennas. In such diffractive coupling conditions, the diffracted waves act to counter the radiation damping of the single nanoantenna response, leading to significant improvements in field enhancement well beyond what is achievable by simply tuning their geometries [75].

In view of applying diffractive coupling to enhanced ultrafast spectroscopy, we investigate the spectral/temporal properties of plasmonic near-fields with FDTD-based numerical simulations. In our simulations, the dimensions of each nanoantenna (length: 1200 nm, width: 200 nm, height: 30 nm) and the arrangement period along the longer axis ($d_x = 2.0\ \mu m$) are fixed, whereas the arrangement period along the shorter axis d_y is varied (Fig. 2.12a).

The calculated field enhancement spectra for varied d_y show crests along the diffractive coupling condition of $d_c = m(\lambda/n)$, where m is a positive integer, λ is the wavelength, and n is the refractive index of the substrate (Fig. 2.12b). Although the isolated nanoantenna exhibits a peak enhancement of 40 at 2000 cm^{-1} and a fwhm bandwidth of 560 cm^{-1}, the nanoantenna arrays of $d_y = 3.2\ \mu m$ (nearly the diffractive coupling condition) exhibit a larger peak enhancement of 80 and a narrower bandwidth of 180 cm^{-1} (Fig. 2.12c).

Figure 2.12d shows the temporal waveforms of the incident electric field of 100-fs duration, the enhanced near-field for the isolated nanoantenna, and that for the nanoantenna arrays of $d_y = 3.2\ \mu m$. The peak electric field is found to be enhanced by 40 times for the isolated nanoantenna and 70 times for the nanoantenna arrays. Such a large enhancement is accompanied by a slight elongation of the temporal duration (120 fs for the isolated nanoantenna and 140 fs for the nanoantenna arrays) due to the spectral filtering effect.

From these results, we can conclude that diffractive coupling is beneficial for higher field enhancements and thus higher sensitivity but sacrifices the time resolution and the available bandwidth. In practice, one should choose adequate d_y to balance the magnitude and bandwidth of the field enhancement, according to the required time resolution (the typical time scale for condensed-phase molecular dynamics varies from tens of femtoseconds to tens of picoseconds).

2.4.4 Linear Spectroscopy of Antenna–Molecule System

Rod-shaped gold nanoantenna arrays are patterned into 2D arrays of rectangular lattices ($d_x = 1.9\ \mu m$, $d_y = 3.3\ \mu m$) on a CaF_2 substrate, whose SEM image is shown in the inset of Fig. 2.13a. On this device, a $W(CO)_6$-dispersed PMMA film of 80-nm thickness is formed. Our target mode is the triply degenerate anti-symmetric CO stretching vibration (T_{1u} mode) of $W(CO)_6$.

As shown in the upper panel of Fig. 2.13a, the extinction spectrum for the polymer film deposited directly on a CaF_2 substrate has two absorption peaks (fwhm bandwidth) at 1976 cm^{-1} (19 cm^{-1}) for the T_{1u} mode of $W(CO)_6$ and at 1731 cm^{-1} (14

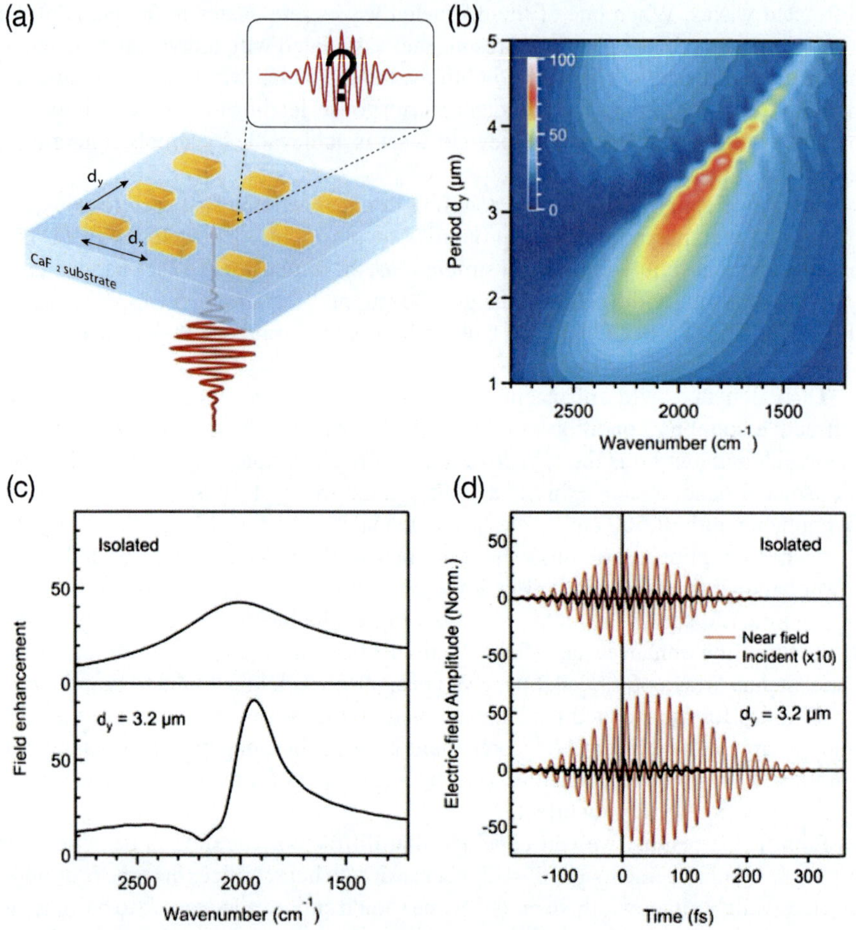

Fig. 2.12 **a** Sketch of the nanoantenna arrays in the numerical simulations. **b** Calculated field enhancement of gold nanoantenna arrays for varied d_y. **c** Field enhancement spectra for $d_y = 3.2$ μm and infinity (isolated). **d** Temporal waveforms of the incident electric field of 100-fs duration, enhanced near-fields for the isolated nanoantenna and for the nanoantenna arrays of $d_y = 3.2$ μm. Adapted from [27]. Copyright ©2017, OSA Publishing

cm^{-1}) for the CO stretching mode of PMMA. In contrast, the extinction spectrum for the polymer film deposited on the nanoantenna arrays (the lower panel in Fig. 2.13a) shows that the broad peak of the plasmonic resonance is modulated at the resonance frequencies of the two vibrational modes, with spectral features characteristic of Fano resonance.

These features are caused by coupling to the molecular vibration in the following manner (see Fig. 2.13b). The incident field excites the plasmon mode (process I), which in turn excites the molecular vibration via near-field coupling (process

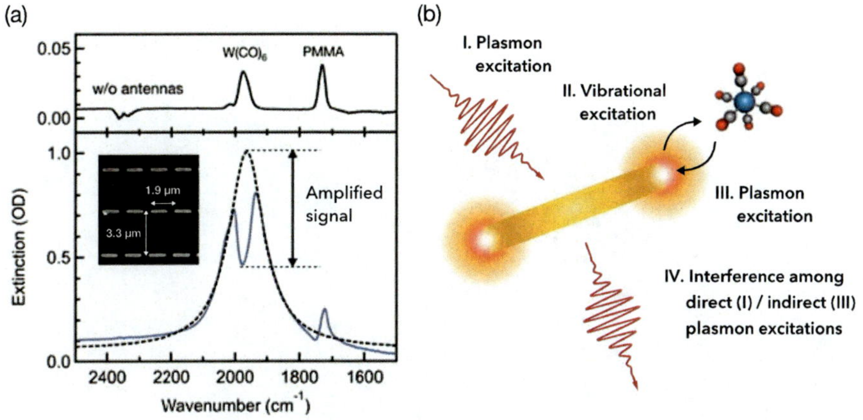

Fig. 2.13 **a** The *upper panel* shows the extinction spectrum for the 80-nm-thick $W(CO)_6$:PMMA film deposited directly on a CaF_2 substrate. The *lower panel* shows that for the diffractively coupled nanoantenna arrays covered (*solid line*) and uncovered (*dashed line*) with the polymer film. The SEM image of the nanoantennas is shown in the *inset*. Each nanoantenna is 1.45 μm long, 0.2 μm wide, and 0.03 μm high. The size of the array unit cell is 1.9 μm × 3.3 μm. **b** Schematic of the antenna-molecule near-field coupling. Adapted from [27]. Copyright ©2017, OSA Publishing

II). Next, the molecular vibrational excitation acts back to excite the plasmon mode (process III). In this manner, coherent dipole excitation is continuously exchanged between the nanoantenna and the molecules until it is damped. Consequently, the modification in the antenna extinction, resulting from such interference among direct/indirect plasmon excitations (process IV), corresponds to the amplified vibrational signal [76, 77].

2.4.5 *Pump-Probe Spectroscopy of Antenna–Molecule System*

We perform single-color IR pump-probe spectroscopy for the T_{1u} mode of $W(CO)_6$ using the setup described in [27]. The spectra of the pump and probe pulses are centered at 2000 cm^{-1} with a fwhm bandwidth of 180 cm^{-1}. The time resolution of the pump-probe measurements is approximately 100 fs.

Figure 2.14c shows the transient absorbance change spectra at scanned delay times for the $W(CO)_6$-dispersed PMMA film without nanoantennas, measured with a pump energy of 890 nJ. Bleaching at the fundamental transition frequency and induced absorption at the red-shifted frequency appear immediately after excitation and decay monotonically. The bleaching signal is attributed to population depletion of the $v = 0$ ground state and stimulated emission from the $v = 1$ state. The induced absorption at the red-shifted frequency is attributed to excited-state absorption of the $v = 1 \rightarrow 2$ transition.

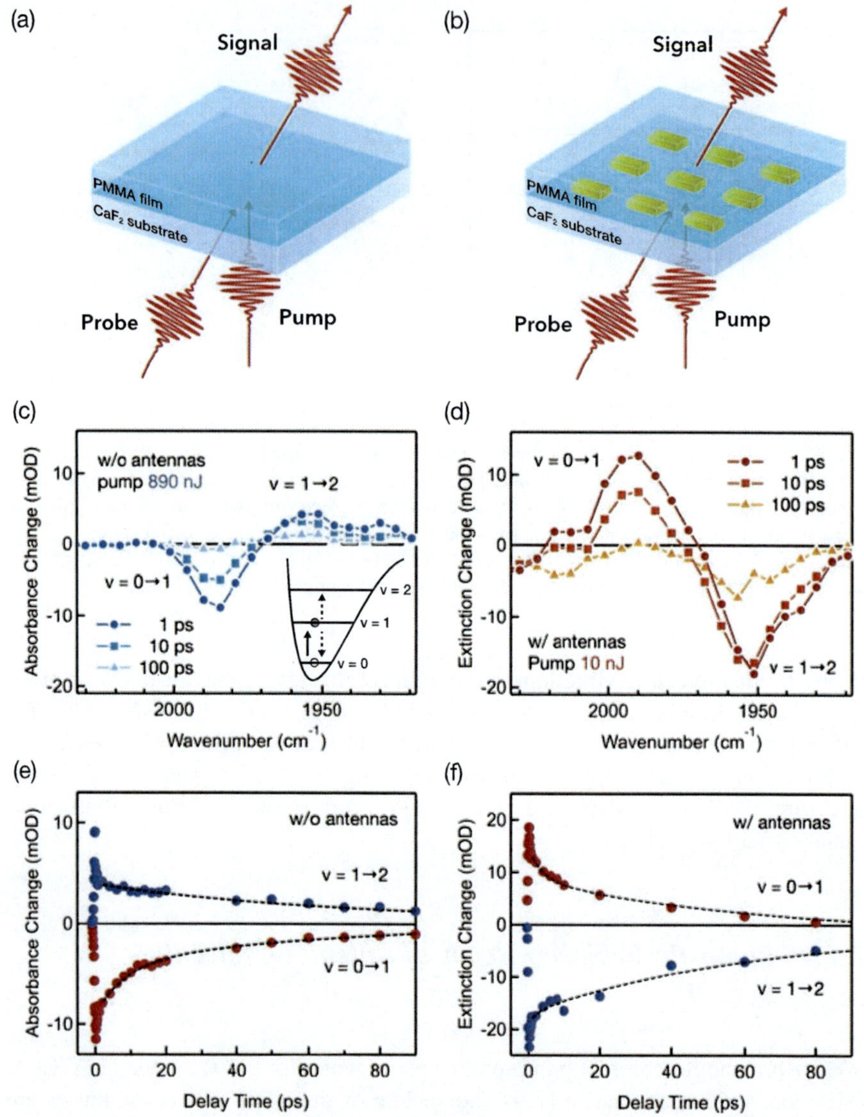

Fig. 2.14 **a**, **b** Schematics of the pump-probe spectroscopy for the $W(CO)_6$-dispersed PMMA film (**a**) without the nanoantennas and (**b**) with the nanoantennas. **c**, **d** Transient extinction change spectra at scanned delay times for the $W(CO)_6$-dispersed PMMA film **c** without the nanoantennas and **d** with the nanoantennas. **e**, **f** Delay scans for the $W(CO)_6$-dispersed PMMA film **e** without the nanoantennas and **f** with the nanoantennas. Adapted from [27]. Copyright ©2017, OSA Publishing

Figure 2.14d shows the transient extinction change spectra at scanned delay times for the $W(CO)_6$-dispersed PMMA film with nanoantennas, measured with a pump energy of 10 nJ. The transient extinction changes display similar behaviors as the typical pump-probe absorbance changes, except for their signs. The inversion of the signs is reasonably explained by the observation of molecular vibrational signals through Fano-like resonant interactions with nanoantennas (Fig. 2.13a). Then, the increased extinction around the fundamental transition frequency is attributed to ground-state bleach($v = 0 \rightarrow 1$) and stimulated emission($v = 1 \rightarrow 0$), whereas the decreased extinction at the red-shifted frequency is attributed to the excited-state absorption($v = 1 \rightarrow 2$).

One can find that the transient extinction changes are comparable between the two measurements shown in Fig. 2.14c, d, despite different pump energies of 890 nJ and 10 nJ, respectively. This indicates that the nonlinear vibrational signal is substantially amplified by the near-field enhancement. From these results, the "spatially-averaged" nonlinear signal enhancement factor, calculated as the extinction change divided by pump energy and then normalized by its value for the bare sample, is estimated to be 850. Furthermore, by assuming that the signals come from only the hot spots near the extremities of the nanoantennas, the "local" nonlinear signal enhancement factor is estimated to be 6 orders of magnitude.

Figure 2.14e, f show the kinetic data at the probe frequencies of bleach (red circle) and induced absorption (blue circle) for the $W(CO)_6$-dispersed PMMA film with/without the nanoantennas, respectively. The global fitting analyses extract almost identical relaxation times of 60–70 ps for the two cases. This fact indicates that antenna–molecule near-field coupling enables nonlinear vibrational spectroscopy with substantially increased signal sensitivity and only minor modifications to molecular vibrational dynamics.

2.4.6 Summary

We have demonstrated surface-enhanced nonlinear vibrational spectroscopy using diffractively coupled gold nanoantenna arrays. The IR pump-probe spectroscopy measurements on a $W(CO)_6$-dispersed PMMA film elucidate that the diffractive coupling condition achieves a local signal enhancement of more than 6 orders of magnitude. Furthermore, the observed vibrational dynamics are almost identical regardless of the signal enhancements by nanoantenna arrays.

Such substantial signal enhancement should be useful for characterizing the structures and dynamics of minute-volume molecules and for the detection of vibrational modes with smaller transition moments. The demonstrated scheme reduces the required pump energy from the micro-joule to the nano-joule level, paving the way for nonlinear vibrational spectroscopy measurements with compact light sources at the oscillator level [78–80]. Such simplification of the whole system is a critical key for nonlinear vibrational spectroscopy to become widely used. Additionally, the oscillator-level pulsed sources generate pulses at higher repetition rates, which is advantageous for shorter data acquisition times.

Reflection geometry, combined with strong field confinement, discards the background signal originating from the bulk, enabling surface-sensitive measurements [28, 81, 82]. Therefore, the reflection-type method is useful for investigating monolayers and interfaces, which are important in the field of structural biology, photocatalysis, and organic electronics.

Further engineering in the metal nanostructure is promising for even higher sensitivity and new functionality. Examples of structures that could provide exceptionally high field enhancements are nanogap structures and perfect-absorber metamaterials [83]. Recently, Chuntonov and coworkers employed plasmonic gap trimers, which have two orthogonal dipole resonances, and demonstrated significant enhancement of the cross-peak signal in dual-frequency 2D-IR spectroscopy [84, 85].

2.5 Reaction Control via Antenna-Enhanced Vibrational Excitation

Mid-IR ultrashort lasers have great potential not only for nonlinear vibrational spectroscopy but also for active control of chemical reactions at the molecular level [86]. Deposition of sufficient vibrational energy into a specific chemical bond via IR laser excitation can lead to selective bond cleavage while leaving the surrounding molecules undisturbed. Such molecular surgery has been achieved for gas-phase molecules [87–89], yet attempts at realizing it in condensed phases, which are essential for practical chemical synthesis, have been hampered so far by poor vibrational excitation efficiency [90–93]. Recently, we utilized the plasmonic field enhancement of mid-IR pulses to strongly drive molecular vibrations and thereby realized bond-breaking of condensed-phase molecules for the first time [29] (Fig. 2.15). This section reviews the historical perspective of vibrationally mediated reaction control and our experimental results.

2.5.1 *Vibrational Control of Chemical Reactions*

Many chemical reactions are driven by thermal activation at the electronic ground state. In such thermally driven reactions, the induced thermal energy populates molecular vibrations participating in the reaction coordinate (RC) and thus propels the reactants over the activation energy barrier to the products. However, heating the molecules results in the excitation of all vibrational modes according to the Boltzmann distribution, which promotes not only desired reactions but also undesired reactions.

Strong excitation of key reagent vibrations is a potentially powerful strategy for bond-selective control of chemical reactions. Upon direct excitation of the RC or by energy relaxation into the RC with vibrational energy above the activation energy barrier, a specific reaction can be promoted while minimizing unwanted side products. Vibrational motions can be activated through electronic excitation, but this scheme is restricted to modes accessible by Franck–Condon transitions and excited-state conical intersections. In contrast, IR excitation of molecular vibrations in electronic ground states may be a more versatile technique because of its much broader accessibility to the potential energy surface.

Mid-IR ultrafast laser pulses have enabled multi-quantum vibrational excitation or vibrational ladder climbing (VLC) and thus opened a way to bond-selective control of chemical reactions in the electronic ground states. Molecular bond-breaking mediated by VLC has been realized for gas-phase molecules [87–89], and similar VLC has been performed on liquid-phase molecules as well [90–93]. Down-chirping has proven useful for efficient VLC [89, 90, 92], and arbitrary pulse-shaping was exploited for selective populations of excited vibrational levels [93]. However, attempts to control ground-state molecular dissociation in liquid phases have been unsuccessful to date because rapid vibrational relaxation disturbs the deposition of sufficient energy in a specific mode [90–93].

One promising approach to resolve these issues is to combine ultrafast laser technology with plasmonics in the mid-IR range. As described in Sect. 2.2, plasmonics takes advantage of the coupling of light to metal electrons, enabling subwavelength localization and strong field enhancements and enhanced light–matter interactions. Since VLC is a higher order nonlinear process, the plasmonic field enhancement may play a key role in accelerating VLC to enable sufficient energy deposition on a timescale faster than vibrational relaxation.

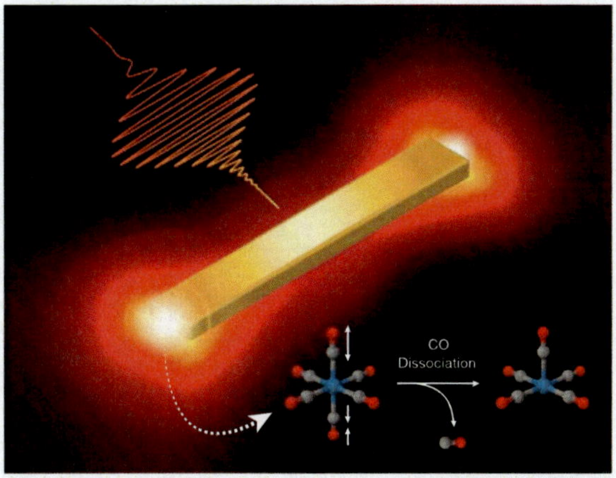

Fig. 2.15 Schematic of reaction control via antenna-enhanced vibrational excitation. The down-chirped mid-IR pulse, enhanced with plasmon excitation of the gold nanoantenna, strongly excites the carbonyl stretch vibration of $W(CO)_6$ molecules, resulting in dissociation of the W-C bond. Adapted from [29]. Copyright ©2019, Nature Publishing Group

2.5.2 *Vibrational Ladder Climbing with Down-Chirped Pulses*

As described above, laser excitation of molecules into a specific higher lying vibrational state is a controllable way to accelerate a desired reaction. Direct excitation of such a higher lying state by single-photon absorption is inefficient because of the small transition dipole moment between the initial and final state. Thus, stepwise excitation of a sequence of vibrational states is a more promising route. Indeed, these transition frequencies decrease while climbing the ladder due to the molecular anharmonicity. Therefore, the climbing efficiency can be increased dramatically when using a negatively chirped pulse so that its high-frequency components, in reso-

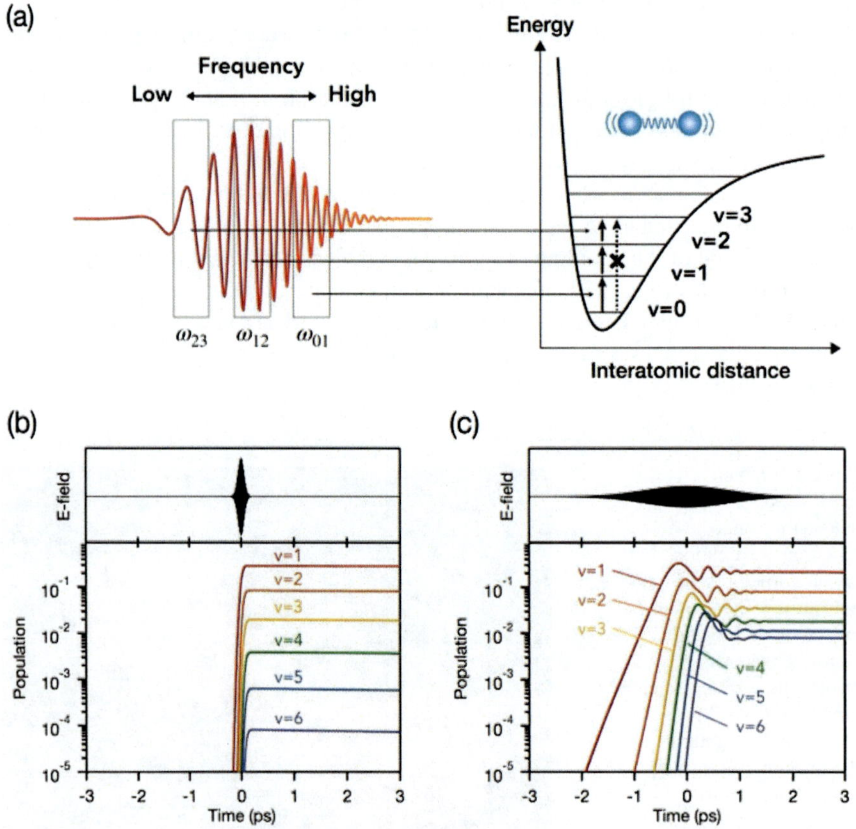

Fig. 2.16 **a** A schematic of vibrational ladder climbing by a down-chirped pulse. **b**, **c** Calculated time evolution of vibrational populations (**b**) for FTL pulse excitation and **c** for down-chirped pulse excitations. The respective electric-field waveforms of the incident pulses are represented in the *upper panels*

nance with the lower transitions of the ladder, precede its low-frequency components, in resonance with the upper transitions of the ladder (Fig. 2.16a).

Figure 2.16b, c show the calculated time evolution of vibrational populations of the T_{1u} CO stretching mode of $W(CO)_6$ excited by Fourier transform limited (FTL) and down-chirped (the group delay dispersion of $-24000\ fs^2$) pulses, respectively (see Supplementary Information of [29] for the detailed description of the calculations). In the case of FTL pulse excitation, the vibrational populations grow simultaneously within the pulse duration. On the other hand, in the case of down-chirped pulse excitation, the vibrational population grows sequentially, resulting in efficiently raising the vibrational population up to the higher lying states.

2.5.3 Antenna-Enhanced Vibrational Ladder Climbing

Our sample is a $W(CO)_6$/*n*-hexane solution held between two CaF_2 windows (Fig. 2.17a). The target mode is the T_{1u} CO stretching mode of $W(CO)_6$, which exhibits an absorption at a center frequency of 1983 cm^{-1} with a fwhm linewidth of 3 cm^{-1} (black line in Fig. 2.17c). The gold nanoantenna arrays are fabricated on one of the two CaF_2 windows, whose SEM image is shown in Fig. 2.17b. The reflectance spectrum for the sample in Fig. 2.17c shows the broad peak of the plasmonic resonance, strongly modified at the resonance frequency of the T_{1u} mode with spectral features characteristic of Fano resonance (see Sect. 2.4.4 for a description of its mechanism).

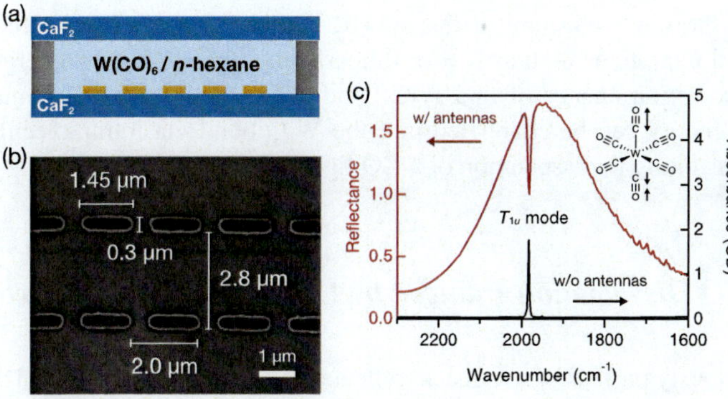

Fig. 2.17 **a** A schematic of the sample comprising two CaF_2 windows separated by a Teflon spacer and filled with $W(CO)_6$/*n*-hexane solution. **b** SEM image of the gold nanoantenna arrays. Each nanoantenna is 1.45 μm long, 0.3 μm wide, and 0.1 μm high. The size of the array unit cell is 2.0 μm × 2.8 μm. **c** The measured reflectance for the nanoantenna arrays immersed in 20 mM $W(CO)_6$/*n*-hexane solution (*red*) and the absorption spectrum for 3 mM $W(CO)_6$/*n*-hexane solution (*black*). These figures are reprinted with permission from [29]. Copyright ©2019, Nature Publishing Group

We perform mid-IR pump-probe spectroscopy for our sample to probe VLC in the T_{1u} mode using mid-IR pulses with a temporal duration of 100 fs and a center frequency of 1930 cm^{-1}. In our experiments, mid-IR pump pulses are locally enhanced with plasmon excitation to strongly drive the T_{1u} mode, and the resulting vibrational population distribution is sensitively probed as the pump-induced reflectance changes.

As a reference experiment, we measure the transient absorption of the uncoupled $W(CO)_6$/n-hexane solution (3 mM) in a transmission geometry using nearly FTL pump pulses with a fluence of 0.2 mJ/cm^2. The transient absorbance change of the uncoupled molecules in Fig. 2.18a shows a negative absorption feature at the fundamental T_{1u} mode frequency (1982 cm^{-1}), attributed to the ground-state bleach ($v = 0 \rightarrow 1$) and the stimulated emission ($v = 1 \rightarrow 0$). They also exhibit a positive absorption feature at 1967 cm^{-1}, attributed to the excited-state absorption ($v = 1 \rightarrow 2$).

Next, we measure the transient reflectance of the antenna-coupled $W(CO)_6$/n-hexane solution (20 mM) in a reflection geometry using nearly FTL pump pulses with a fluence of 0.2 mJ/cm^2. The transient reflectance change of the antenna-coupled molecules in Fig. 2.18b shows a positive reflectance change at 1982 cm^{-1}, attributed to the ground-state bleach ($v = 0 \rightarrow 1$) and the stimulated emission ($v = 1 \rightarrow 0$). They also exhibit negative reflectance changes, attributed to the excited-state absorptions: $v = 1 \rightarrow 2$ at 1967 cm^{-1}, $v = 2 \rightarrow 3$ at 1953 cm^{-1}, $v = 3 \rightarrow 4$ at 1931 cm^{-1}, $v = 4 \rightarrow 5$ at 1908 cm^{-1}, and $v = 5 \rightarrow 6$ at 1889 cm^{-1}. When the pump pulses are down-chirped with the group delay dispersion of -24000 fs^2, we observe an additional signal at 1866 cm^{-1}, attributed to the excited-state absorption of the $v = 6 \rightarrow 7$ transition (Fig. 2.18c).

From these results, it can be reasonably concluded that the near-field enhancement as well as the down-chirping of the mid-IR pump pulses plays essential roles in realizing VLC up to as high as $v = 6$. Remarkably, this vibrational energy is close to the dissociation energy of one W-C bond. This suggests that the multiphoton excitation energy can be channeled into the W-C bond via intramolecular mode-coupling, resulting in dissociation of a CO ligand.

2.5.4 CO Dissociation Induced by Chirp-Pulsed Near-Fields

We continuously measure the transient reflectance of the antenna-coupled molecules for an hour while irradiating the down-chirped pump pulses. Figure 2.19a shows the transient reflectance change measured before and after an hour of pump irradiation. Surprisingly, the 1 h of irradiation induces a change in the transient reflectance; i.e., there emerges an additional positive signal in the spectral range of 1910–1950 cm^{-1}. From the transient spectrum for 60 min, subtracted by the one for 0 min, it is clearly observed that the new band exhibits a center frequency of 1930 cm^{-1} and a fwhm linewidth of 22 cm^{-1}.

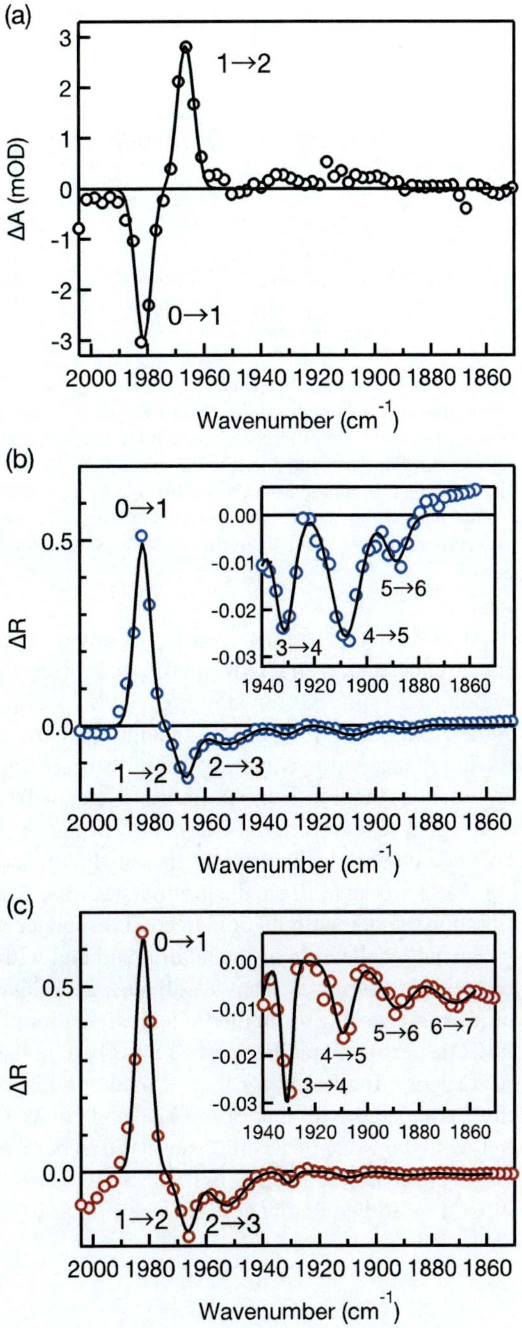

Fig. 2.18 **a** The transient absorbance change of uncoupled $W(CO)_6$ at a 2-ps time delay after excitation with nearly FTL pulses. **b**, **c** The transient reflectance change of coupled $W(CO)_6$ at a 2-ps time delay after excitation: **b** with nearly FTL pulses, and **c** with down-chirped pulses. Each *black line* represents a linear combination of multiple Gaussian functions fitted to the measured absorbance/reflectance change. These figures are reprinted with permission from [29]. Copyright ©2019, Nature Publishing Group

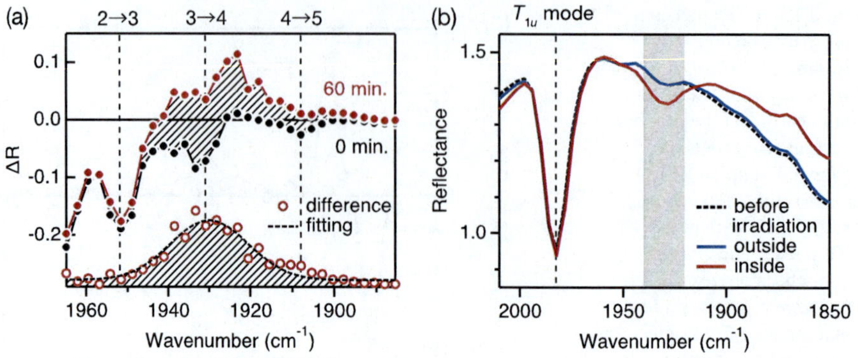

Fig. 2.19 **a** Transient reflectance change at a 2-ps time delay for the antenna-coupled $W(CO)_6$ before (*black filled markers*) and after (*red filled markers*) the 1 h of irradiation. The *red open markers* represent the difference between the two reflectance changes, and the *black dashed line* represents a Gaussian function fitted to the data. **b** The linear reflectance spectrum measured after 1 h of repeated irradiation for the area inside (*red*) and outside (*blue*) the irradiation beam spot. A *black dashed line* denotes the spectrum measured before irradiation. These figures are reprinted with permission from [29]. Copyright ©2019, Nature Publishing Group

After the 1 h of irradiation with the down-chirped pump pulses, we measure the linear reflectance with FT-IR microscope for the sample area inside/outside the irradiation beam spot, as shown in Fig. 2.19b. In addition to the original absorption dip of the T_{1u} mode at 1982 cm^{-1}, an additional absorption dip is observed for the area inside the irradiation beam spot (not observed for the area outside the beam spot). Importantly, this dip exhibits the same center frequency and linewidth as the positive signal observed in the transient reflectance change with the pump irradiation. This coincidence indicates that the newly appearing pump-probe signal shown in Fig. 2.19a originates from the ground-state bleach and the stimulated emission of the vibrational mode with its fundamental resonance at 1930 cm^{-1}.

Spatial localization of the additional band within the irradiation beam spot suggests that it originates from adsorbates, i.e., CO or $W(CO)_5$ adsorbed on the gold surfaces. According to our theoretical calculations based on density functional theory, $W(CO)_5$ becomes stabilized when adsorbed on the gold surface, and the fundamental transition frequency for the A_1 mode of adsorbed $W(CO)_5$ is red-shifted by 45 cm^{-1} compared with that of the T_{1u} mode of $W(CO)_6$. This frequency shift agrees well with the value of 53 cm^{-1} observed in our experiments. From these results, we assign the absorption band observed at 1930 cm^{-1} to the A_1 CO stretching mode of $W(CO)_5$ adsorbed on the gold surface (Fig. 2.20).

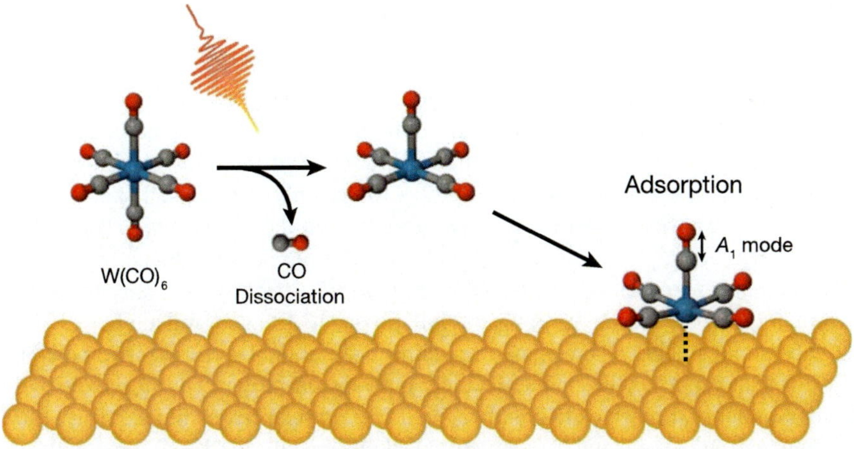

Fig. 2.20 Proposed reaction schemes induced by antenna-enhanced vibrational ladder climbing. Upon multiphoton vibrational excitation, the $W(CO)_6$ molecule loses one CO ligand. CO dissociation is followed by adsorption on the gold surface. This figure is reprinted with permission from [29]. Copyright ©2019, Nature Publishing Group

2.5.5 *Summary*

We have demonstrated successful driving of ground-state dissociation of liquid-phase molecules by employing plasmonic near-fields of chirped mid-IR pulses. Both down-chirping and plasmonic field enhancement play essential roles in raising vibrational populations to higher lying states, realizing bond dissociation. Interestingly, the gold nanoantennas play an additional role in stabilizing the dissociated species prior to recombination. These key roles solve the long-standing issues of poor excitation efficiency and diffusional recombination that have hampered vibration-mediated dissociation of condensed-phase molecules.

The demonstrated scheme should be useful for mode-selective control of not only bond-breaking but also other kinds of chemical reactions. For instance, Heyne and coworkers have recently demonstrated the acceleration of bimolecular reaction, the alcoholysis of urethane, by exciting a vibrational mode of the reactants [94]. Weinstein and coworkers have achieved mode-specific control of the path of light-induced electron transfer at a molecular fork via vibrational excitation [95].

Our scheme can be developed into more powerful and versatile techniques by incorporating arbitrary pulse-shaping, well-designed plasmonic nanostructures, and functions of metal surfaces, paving the way toward controlled ground-state chemistry.

2.6 Summary and Prospects

As presented in the preceding sections, gold nanoantennas efficiently couple femtosecond mid-IR laser pulses into plasmonic near-fields to substantially enhance ultrafast light–matter interaction at the nanoscale. And thus, these gold nanoantennas have brought about ultrasensitivity onto nonlinear vibrational spectroscopy (Sect. 2.4), and have realized an vibrationally mediated chemical reaction control in a liquid phase (Sect. 2.5). These successful demonstrations illustrate that *ultrafast mid-IR plasmonics form a novel platform for vibrational spectroscopy and vibrational quantum control*, allowing us to view future prospects from several angles.

Substantial signal enhancement in nonlinear vibrational spectroscopy, including 2D-IR spectroscopy, will enable detailed analyses on protein conformations with fewer number of molecules to greatly benefit protein science and biochemistry [16].

Engineering in the metal nanostructure is promising for increased sensitivity and new functionality. For instance, nanogap structures and perfect-absorber metamaterials provide exceptionally high field enhancement [83], whereas plasmonic gap trimers provide a new degree in polarization control of near-fields that allows sequential excitation of multiple vibrational modes whose transition dipole moments may have different orientations [84, 85].

Near-field enhancement at a surface brings about another benefit of *surface sensitivity* to spectroscopy. Especially, antenna-enhanced spectroscopy exploiting reflection geometry (described in Sect. 2.4) provides better surface sensitivity and therefore would be suited for the investigation of monolayers and interfaces, which are important in the field of structural biology, electrochemistry and organic electronics [81]. Although gold and other noble metals exhibit superior field enhancement, the plasmonic near-fields are only weakly bound to metal surfaces. For improved surface sensitivity, other materials with plasma frequencies in the mid-IR range are preferable. Promising candidates are silicides (NiSi and TiSi), doped semiconductors (Si, Ge, and InAs), and graphene [32]. Graphene offers tunable plasma frequency in the mid-IR to THz ranges and strong subwavelength confinement due to its unique band structure [96, 97].

Another benefit of signal enhancement is the reduction in the required pulse energy from the micro-joule to the nano-joule level, which paves the way for nonlinear vibrational spectroscopy measurements with compact light sources of oscillator level, typically operated at the MHz repetition rate [78–80]. This will shorten the acquisition time as well as improve the signal-to-noise ratio to accelerate the wide use of nonlinear vibrational spectroscopy.

The quantum control scheme demonstrated in Sect. 2.5 will be useful for selective control of various kinds of chemical reactions, including an intramolecular bond-breaking reaction, a bimolecular reaction of the alcoholysis of urethane [94], controlled electron transfer at a molecular fork [95], etc. The scheme can be developed into more powerful and versatile techniques by incorporating engineered plasmonic nanostructures, arbitrary pulse-shaping, and functions of metal surfaces, making innovative advance in the long-standing dream of controlled ground-state chemistry.

In parallel to these plasmon-enhanced nonlinear vibrational processes, ultrafast plasmonics play essential roles in the emerging field of strong-field nanoscale science [98, 99]. In this field, ultrashort mid-IR pulses are beneficial because of their ability to provide large ponderomotive energy and to apply strong fields onto condensed matter without materials damage [100]. Here, nanoscale localization of an ultrashort-pulsed field enables coherent control of free-electron wavepackets and photocurrents on nanometer scale. The two fields that share common technologies of ultrafast mid-IR plasmonics may be developed into new coherent control schemes that treat both electronic and vibrational degrees of freedom at the mesoscopic scale.

The authors thank Fumiya Kusa, Akinobu Takegami, and Shingo Usui at Tokyo University of Agriculture and Technology and Atsunori Sakurai, Kei Murata, Kazuyuki Ishii, Kotaro Imasaka, Govind Dayal Singh, Shuta Kitade, and Nobuyoshi Hiramatsu at the University of Tokyo for their contributions to the presented works. The authors appreciate Claus Ropers, Georg Herink, Katharina Echternkamp, Lara Wimmer, and Murat Sivis at Göttingen University for the collaboration on the photoemission experiment described in Sect. 2.3 and for stimulating discussion on the relevant topics. S. A. acknowledges the support of Japan Society for the Promotion of Science (20H02651, 20K20556, 18K19030, 26600113) and Ministry of Education, Culture, Sports, Science and Technology (Q-LEAP JPMXS0118068681); I. M. acknowledges the support of Grants-in-Aid for JSPS Research Fellow (No. JP18J11212).

References

1. A.H. Zewail, Pure Appl. Chem. **72**, 2219 (2000)
2. S.A. Rice, Nature **409**, 422 (2001)
3. M. Shapiro, P. Brumer, Rep. Prog. Phys. **66**, 859 (2003)
4. B. Kohler, J.L. Krause, F. Raksi, K.R. Wilson, V.V. Yakovlev, R.M. Whitnell, Y. Yan, Acc. Chem. Res. **28**, 133 (1995)
5. R.J. Gordon, S.A. Rice, Annu. Rev. Phys. Chem. **48**, 601 (1997)
6. H. Rabitz, R. de Vivie-Riedle, M. Motzkus, K. Kompa, Science **288**, 824 (2000)
7. M.I. Stockman, S.V. Faleev, D.J. Bergman, Phys. Rev. Lett. **88**, 067402 (2002)
8. M. Aeschlimann, M. Bauer, D. Bayer, T. Brixner, F. J. GarcAa de Abajo, W. Pfeiffer, M. Rohmer, C. Spindler, F. Steeb: Nature **446**, 301 (2007)
9. P. Vasa, C. Ropers, R. Pomraenke, C. Lienau, Laser & Photon. Rev. **3**, 483 (2009)
10. M. Aeschlimann, M. Bauer, D. Bayer, T. Brixner, S. Cunovic, F. Dimler, A. Fischer, W. Pfeiffer, M. Rohmer, C. Schneider, F. Steeb, C. StrAŒber, D.V. Voronine, Proc. Nat. Acad. Sci. **107**, 5329 (2010)
11. S. Berweger, J.M. Atkin, X.G. Xu, R.L. Olmon, M.B. Raschke, Nano Lett. **11**, 4309 (2011)
12. D. Brinks, M. Castro-Lopez, R. Hildner, N.F. van Hulst, Proc. Nat. Acad. Sci. **110**, 18386 (2013)
13. M.D. Fayer, *Ultrafast Infrared Vibrational Spectroscopy* (CRC Press, 2013)
14. P. Hamm, M.T. Zanni, *Concepts and Methods of 2D Infrared Spectroscopy* (Cambridge University Press, 2011)
15. D. Laage, T. Elsaesser, J.T. Hynes, Chem. Rev. **117**, 10694 (2017)
16. A. Ghosh, J.S. Ostrander, M.T. Zanni, Chem. Rev. **117**, 10726 (2017)
17. J.C. Polanyi, Acc. Chem. Res. **5**, 161 (1972)

18. N. Doslić, K. Sundermann, L. González, O. Mó, J. Giraud-Girard, O. Kühn, Phys. Chem. Chem. Phys. **1**, 1249 (1999)
19. Y. Ohtsuki, M. Sugawara, H. Kono, Y. Fujimura, Bull. Chem. Soc. Jpn **74**, 1167 (2001)
20. M. Artamonov, T.S. Ho, H. Rabitz, Chem. Phys. **328**, 147 (2006)
21. C. Gollub, M. Kowalewski, S. Thallmaira, R. de Vivie-Riedle, Phys. Chem. Chem. Phys. **12**, 15780 (2010)
22. F. Kusa, S. Ashihara, J. Appl. Phys. **116**, 153103 (2014)
23. N. Hiramatsu, F. Kusa, K. Imasaka, I. Morichika, A. Takegami, S. Ashihara, J. Appl. Phys. **120**, 173103 (2016)
24. S. Usui, S. Kitade, I. Morichika, K. Kohmura, F. Kusa, S. Ashihara, J. Phys. Chem. C **121**, 26000 (2017)
25. F. Kusa, K.E. Echternkamp, G. Herink, C. Ropers, S. Ashihara, AIP Adv. **5**, 077138 (2015)
26. K. Imasaka, T. Kaji, T. Shimura, S. Ashihara, Opt. Express **26**, 21364 (2018)
27. F. Kusa, I. Morichika, A. Takegami, S. Ashihara, Opt. Express **25**, 12896 (2017)
28. I. Morichika, F. Kusa, A. Takegami, A. Sakurai, S. Ashihara, J. Phys. Chem. C **121**, 11643 (2017)
29. I. Morichika, K. Murata, A. Sakurai, K. Ishii, S. Ashihara, Nat. Commun. **10**, 3893 (2019)
30. R. Stanley, Nat. Photonics **6**, 409 (2012)
31. S. Law, V. Podolskiy, D. Wasserman, Nanophotonics **2**, 103 (2013)
32. Y. Zhong, S.D. Malagari, T. Hamilton, D. Wasserman, J. Nanophoton. **9**, 093791 (2015)
33. A. Hartstein, J.R. Kirtley, J.C. Tsang, Phys. Rev. Lett. **45**, 201 (1980)
34. F. Neubrech, A. Pucci, T.W. Cornelius, S. Karim, A. GarcAa-Etxarri, J. Aizpurua, Phys. Rev. Lett. **101**, 157403 (2008)
35. R. Adato, A. A. Yanik, J. J. Amsden, D. L. Kaplan, F. G. Omenetto, M. K. Hong, S. Erramilli, H. Altug: Proc. Nat. Acad. Sci. USA textbf106, 19227 (2009)
36. C.V. Hoang, M. Oyama, O. Saito, M. Aono, T. Nagao, Sci. Rep. **3**, 1175 (2013)
37. F. Neubrech, C. Huck, K. Weber, A. Pucci, H. Giessen, Chem. Rev. **117**, 5110 (2017)
38. X. Yang, Z. Sun, T. Low, H. Hu, X. Guo, F.J.G. de Abajo, P. Avouris, Q. Dai, Adv. Mater. **30**, 1704896 (2018)
39. R. Kas, O. Ayemoba, N.J. Firet, J. Middelkoop, W.A. Smith, A. Cuesta, Chem. Phys. Phys. Chem. **133**, 12881 (2019)
40. J. Cleary, R. Peale, D. Shelton, G. Boreman, R. Soref, W. Buchwald: MRS Proc. 1133-AA10-03 (2008)
41. A. Tittl, A. John-Herpin, A. Leitis, E.R. Arvelo, H. Altug, Angew. Chem. Int. Ed. **58**, 14810 (2019)
42. F. Kusunoki, J. Takahara, I. Kobayashi, Electron. Lett. **39**, 23 (2003)
43. T.W. Ebbesen, C. Genet, S.I. Bozhevolnyi, Phys. Today **61**, 44 (2008)
44. R. Soref, R.E. Peale, W. Buchwald, Opt. Express **16**, 6507 (2008)
45. R.L. Olmon, B. Slovick, T.W. Johnson, D. Shelton, S.-H. Oh, G.D. Boreman, M.B. Raschke, Phys. Rev. B **86**, 235147 (2012)
46. S. Law, L. Yu, D. Wasserman, J. Vac. Sci. Technol. B **31**, 03C121 (2013)
47. A.V. Zayats, Nature **495**, S7 (2013)
48. J. Zuloaga, P. Nordlander, Nano Lett. **11**, 1280 (2011)
49. P. Hommelhoff, Y. Sortais, A. Aghajani-Talesh, M.A. Kasevich, Phys. Rev. Lett. **96**, 077401 (2006)
50. C. Ropers, D.R. Solli, C.P. Schulz, C. Lienau, T. Elsaesser, Phys. Rev. Lett. **98**, 043907 (2007)
51. R. Bormann, M. Gulde, A. Weismann, S.V. Yalunin, C. Ropers, Phys. Rev. Lett. **105**, 147601 (2010)
52. M. Schenk, M. Kruger, P. Hommelhoff, Phys. Rev. Lett. **105**, 257601 (2010)
53. M. Krüger, M. Schenk, P. Hommelhoff, Nature **475**, 78 (2011)
54. G. Herink, D.R. Solli, M. Gulde, C. Ropers, Nature **483**, 190 (2012)
55. V. Schweikhard, A. Grubisic, T.A. Baker, D.J. Nesbitt, J. Phys. Chem. C **115**, 83 (2011)
56. A. Grubisic, V. Schweikhard, T.A. Baker, D.J. Nesbitt, Acs Nano **7**, 87 (2013)

57. P. Dombi, A. Horl, P. Racz, I. Marton, A. Trugler, J.R. Krenn, U. Hohenester, Nano Lett. **13**, 674 (2013)
58. S.M. Teichmann, P. Racz, M.F. Ciappina, J.A. Perez-Hernandez, A. Thai, J. Fekete, A.Y. Elezzabi, L. Veisz, J. Biegert, P. Dombi, Sci. Rep. **5**, 7584 (2015)
59. R.G. Hobbs, Y. Yang, A. Fallahi, P.D. Keathley, E. De Leo, F.X. Kartner, W.S. Graves, K.K. Berggren, Acs Nano **8**, 11474 (2014)
60. M. Gulde, S. Schweda, G. Storeck, M. Maiti, H.K. Yu, A.M. Wodtke, S. Schafer, C. Ropers, Science **345**, 200 (2014)
61. A. Mustonen, P. Beaud, E. Kirk, T. Feurer, S. Tsujino, Appl. Phys. Lett. **99**, 103504 (2011)
62. D.S. Yang, O.F. Mohammed, A.H. Zewail, Proc. Natl. Acad. Sci. U. S. A. **107**, 14993 (2010)
63. W.S. Graves, F.X. Kartner, D.E. Moncton, P. Piot, Phys. Rev. Lett. **108**, 263904 (2012)
64. L. Wimmer, G. Herink, D.R. Solli, S.V. Yalunin, K.E. Echternkamp, C. Ropers, Nat. Phys. **10**, 432 (2014)
65. M.I. Stockman, M.F. Kling, U. Kleineberg, F. Krausz, Nat. Photon. **1**, 539 (2007)
66. G. Herink, L. Wimmer, C. Ropers, New J. Phys. **16**, 123005 (2014)
67. M. Krüger, M. Schenk, M. Förster, P. Hommelhoff, J. Phys. B: At. Mol. Opt. Phys. **45**, 074006 (2012)
68. L.V. Keldysh, Sov. Phys. Jetp-Ussr. **20**, 1307 (1965)
69. R. H. Fowler, L. W. Nordheim: Proc. R. Soc. London, Ser **A119**, 173 (1928)
70. C.A. Spindt, I. Brodie, L. Humphrey, E.R. Westerberg, J. Appl. Phys. **47**, 5248 (1976)
71. G.G. Paulus, W. Becker, H. Walther, Phys Rev A **52**, 4043 (1995)
72. P. Alonso-Gonzalez, P. Albella, F. Neubrech, C. Huck, J. Chen, F. Golmar, F. Casanova, L.E. Hueso, A. Pucci, J. Aizpurua, R. Hillenbrand, Phys. Rev. Lett. **110**, 203902 (2013)
73. P.M. Donaldson, P. Hamm, Angew. Chem. Int. Ed. **52**, 634 (2013)
74. O. Selig, R. Siffels, Y.L.A. Rezus, Phys. Rev. Lett. **114**, 233004 (2015)
75. V.G. Kravets, A.V. Kabashin, W.L. Barnes, A.N. Grigorenko, Chem. Rev. **118**, 5912 (2018)
76. I. Morichika, F. Kusa, A. Takegami, and S. Ashihara: Proc. of SPIE **10252**, Optical Manipulation Conference, 1025215 (2017)
77. I. Morichika, A. Sakurai, S. Ashihara, EPJ Web of Conf. **205**, 03002 (2019)
78. J. Ma, Z. Qin, G. Xie, L. Qian, D. Tang, Appl. Phys. Rev. **6**, 021317 (2019)
79. D. Okazaki, H. Arai, A. Anisimov, E.I. Kauppinen, S. Chiashi, S. Maruyama, N. Saito, S. Ashihara, Opt. Lett. **44**, 1750 (2019)
80. D. Okazaki, I. Morichika, H. Arai, E.I. Kauppinen, Q. Zhang, A. Anisimov, I. Varjos, S. Chiashi, S. Maruyama, S. Ashihara, Opt. Express **28**, 19997 (2020)
81. J.P. Kraack, P. Hamm, Chem. Rev. **117**, 10623 (2017)
82. M.K. Petti, J.S. Ostrander, V. Saraswat, E.R. Birdsall, K.L. Rich, J.P. Lomont, M.S. Arnold, M.T. Zanni, J. Chem. Phys. **150**, 024707 (2019)
83. Y. Yang, C. Gu, J. Li, Small **15**, e1804177 (2019)
84. B. Cohn, B. Engelman, A. Goldner, L. Chuntonov, J. Phys. Chem. Lett. **9**, 4596 (2018)
85. R.T. Mackin, B. Cohn, B. Engelman, A. Goldner, I.V. Rubtsov, L. Chuntonov, J. Phys. Chem. C **123**, 24731 (2019)
86. K. Heyne, O. Kuhn, J. Am. Chem. Soc. **141**, 11730 (2019)
87. L. Windhorn, T. Witte, J.S. Yeston, D. Proch, M. Motzkus, K.L. Kompa, W. Fuss, Chem. Phys. Lett. **357**, 85 (2002)
88. L. Windhorn, J.S. Yeston, T. Witte, W. Fuss, M. Motzkus, D. Proch, K.-L. Kompa, C.B. Moore, J. Chem. Phys. **119**, 641 (2003)
89. T. Witte, T. Hornung, L. Windhorn, D. Proch, R. de Vivie-Riedle, M. Motzkus, K.L. Kompa, J. Chem. Phys. **118**, 2021 (2003)
90. V.D. Kleiman, S.M. Arrivo, J.S. Melinger, E.J. Heilweil, Chem. Phys. **233**, 207 (1998)
91. T. Witte, J.S. Yeston, M. Motzkus, E.J. Heilweil, K.L. Kompa, Chem. Phys. Lett. **392**, 156 (2003)
92. C. Ventalon, J.M. Fraser, M.H. Vos, A. Alexandrou, J.L. Martin, M. Joffre, Proc. Nat. Ac. Sci. **101**, 13216 (2004)
93. D.B. Strasfeld, S.H. Shim, M.T. Zanni, Phys. Rev. Lett. **99**, 038102 (2007)

94. T. Stensitzki, Y. Yang, V. Kozich, A.A. Ahmed, F. Kössl, O. Kühn, K. Heyne, Nat. Chem. **10**, 126 (2018)
95. M. Delor, S.A. Archer, T. Keane, A.J.H.M. Meijer, I.V. Sazanovich, G.M. Greetham, M. Towrie, J.A. Weinstein, Nat. Chem. **9**, 1099 (2017)
96. A.N. Grigorenko, M. Polini, K.S. Novoselov, Nat. Photon. **6**, 749 (2012)
97. G.X. Ni, A.S. McLeod, Z. Sun, L. Wang, L. Xiong, K.W. Post, S.S. Sunku, B.-Y. Jiang, J. Hone, C.R. Dean, M.M. Fogler, D.N. Basov, Nature **557**, 530 (2018)
98. M.F. Ciappina et al., Rep. Prog. Phys. **80**, 054401 (2017)
99. P. Dombi et al., Rev. Mod. Phys. **92**, 025003 (2020)
100. S. Ghimire, D.A. Reis, Nat. Phys. **15**, 10 (2019)

Chapter 3
Plasmonic Photochemical Water Splitting for Efficient Solar Energy Conversion

Xu Shi and Hiroaki Misawa

Abstract Photochemical water splitting using solar energy has been widely studied as a promising strategy for clean and sustainable energy production. Localized surface plasmon resonance has been proposed to enhance the efficiency of photochemical water splitting based on its novel properties. This chapter presents recent progress in the development of plasmon-induced water splitting, especially photoelectrochemical water splitting. The mechanisms of plasmonic chemical reactions are introduced, and recent works on emerging strategies for efficient plasmonic water splitting are summarized. Finally, the future prospects of plasmonic water splitting are discussed.

3.1 Introduction

Solar energy is an inexhaustible and renewable energy source. After scattering and absorption loss through the atmosphere, the energy that reaches Earth's surface is far greater than the energy consumption worldwide. The developments and applications of solar energy conversion systems such as solar heating [1] photovoltaics [2] and artificial photosynthesis [3, 4] have attracted broadening interest. Although the rapid deployment of solar energy conversion devices has been achieved, solar technologies still suffer from some drawbacks, such as high capital cost, modest conversion efficiency, and intermittency. Among these drawbacks, intermittency is a crucial problem when using solar energy. To address this problem, direct production of fuels (H_2 or carbon-based feedstocks) using solar energy is a valuable strategy for overcoming the intermittency disadvantage.

X. Shi (✉)
Creative Research Institution, Hokkaido University, Sapporo 001-0021, Japan
e-mail: shixu@es.hokudai.ac.jp

H. Misawa
Research Institute for Electronic Science, Hokkaido University, Sapporo 001-0020, Japan
e-mail: misawa@es.hokudai.ac.jp

Centre for Emergent Functional Matter Science, National Yang Ming Chiao Tung University, Hsinchu 30010, Taiwan

T. Yatsui (ed.), *Progress in Nanophotonics 6*, Nano-Optics and Nanophotonics,
https://doi.org/10.1007/978-3-030-71516-8_3

Solar hydrogen production from photochemical water splitting shows promise for meeting the fast-increasing energy consumption and solving the current energy crisis. Photoelectrochemical (PEC) water splitting is capable of producing hydrogen in an efficient and cost-effective way. A tremendous number of studies on PEC cells were triggered by the discovery of water splitting with TiO_2 photoelectrodes under UV irradiation in 1972 [5]. Major international research on PEC water splitting was launched after the oil crisis in 1973. The essential part of a PEC cell is the semiconductor, which is a light-absorbing material that converts solar energy to charge carriers followed by chemical reactions. Metal oxide semiconductors, such as TiO_2, $SrTiO_3$, ZnO, and Fe_2O_3, have been extensively studied as photoelectrodes for water splitting applications to produce hydrogen [6]. TiO_2 has been extensively investigated over the last four decades because of its high photoactivities, low cost, low toxicity, and long-term stability [7, 8]. However, only 4% of the entire solar spectrum can be used to directly drive the photovoltaic and photochemical effects of TiO_2. Several methods have been proposed to improve the photoresponse of TiO_2, including doping [9–11], defect creation [12], and dye sensitization [13]. In addition, a novel approach to solve this problem is the use of plasmonic metal NPs (NPs), which can sensitize wide-bandgap semiconductors to visible light [14, 15]. Metallic NPs showing localized surface plasmon resonance (LSPR) exhibit large absorption/scattering cross-sections and an ultrahigh electromagnetic near field, which can be tailored throughout the visible to near-infrared region by varying their size, shape, and surrounding material [16, 17]. With the development of elaborate nanofabrication technologies, plasmonic metal NPs have been applied across broad research fields, including surface-enhanced Raman scattering [18, 19] sensing [20], nonlinear optics [21, 22], and solar energy conversion [14, 23]. Over recent decades, many studies have also demonstrated plasmon-enhanced photocurrent generation and photochemical water splitting via photoinduced energy or electron transfer from metallic NPs to semiconductors via localized surface plasmon resonance [24–27].

In the following sections, we will present the main mechanisms of plasmonic chemical reactions and the recent works on the emerging strategies for efficient plasmonic water splitting as well as the future prospects of plasmonic water splitting.

3.2 Mechanisms of Plasmonic Enhancement

There are three main mechanisms by which plasmons can enhance photochemical reactions on semiconductors [15, 28, 29]. These mechanisms will be discussed in detail below.

3.2.1 Plasmon-Induced Electron Transfer

The principle of plasmon-induced electron transfer is known as the LSPR sensitization effect, analogous to dye-sensitized solar cells, in which plasmonic metallic NPs are used as a sensitizer instead of dye molecules. The plasmonic optical antenna is promising because its optical resonance wavelength can be easily controlled from the visible to the infrared region by changing its geometry [16]. Research on LSPR-driven electron transfer has focused largely on the Au/TiO_2 system [30–33]. It will be convenient to use Au/TiO_2 as a model to discuss the electron transfer process in this section. When a Au NP is applied on the surface of TiO_2, equilibration of the Fermi levels causes bending of the conduction band of a typical n-type TiO_2 at the interface, and a Schottky barrier is formed between the Au and TiO_2. The height of the Schottky barrier is determined only by the Fermi levels of the Au and the conduction band levels of the TiO_2 and is generally on the order of 1.0 eV [34]. Upon light irradiation, the LSPR in the Au NP drives the electron to generate a coherent oscillation in the Au NP, and the energy is absorbed via either the interband transition or the intraband transition (see Fig. 3.1a). Because of electron-electron scattering, electron energy is redistributed and creates a nonequilibrium distribution within approximately 10 femtoseconds [35, 36]. Furthermore, the energetic electrons with energies larger than the Schottky barrier can transfer from the Au NPs to the conduction band of TiO_2, as indicated by the blue arrow in Fig. 3.1b. This process is also called plasmon-induced indirect charge transfer. Since the light energy used to excite the electron transfer from the Au NPs to TiO_2 depends only on the height of the Schottky barrier, the absorption band edge is thereby unlimited by the bandgap of TiO_2 and can be extended to the visible and near-infrared regions.

Early studies on Au-TiO_2 nanocomposites dated back to 1996, [37] although the mechanism was unclear at that time. Later, a study on plasmon-induced charge separation in Au-TiO_2 was conducted in 2005 [24]. The plasmon-excited electrons in Au were proposed to transfer to the conduction band of the adjacent TiO_2. Hot-electron transfer from Au NPs to TiO_2 has also been studied using femtosecond transient absorption spectroscopy with an IR probe, in which a 40% yield of electron injection from plasmon-excited Au NPs into TiO_2 was observed [25]. Based on electron transfer under plasmon resonance, a photocurrent photodetector that uses Au NPs as an active optical antenna loaded on silicon was also reported [38]. Energetic electron transfer competes with electron relaxation through rapid electron-electron scattering in the conduction band of the Au NPs on time scales of hundreds of femtoseconds [39]. Efficient electron transfer requires interfacial charge separation at the Au/TiO_2 interface on an even faster time scale, which is a major challenge for the implementation of efficient plasmonic devices.

As mentioned above, plasmons decay into a hot electron-hole pair within the metal via electron-electron scattering on time scales of tens to hundreds of femtoseconds followed by the transfer of the electron into adjacent semiconductors. Hot-electron transfer competes with electron relaxation through rapid electron-electron scattering in the metal's conduction band. Since the lifetime of plasmon-induced charge carri-

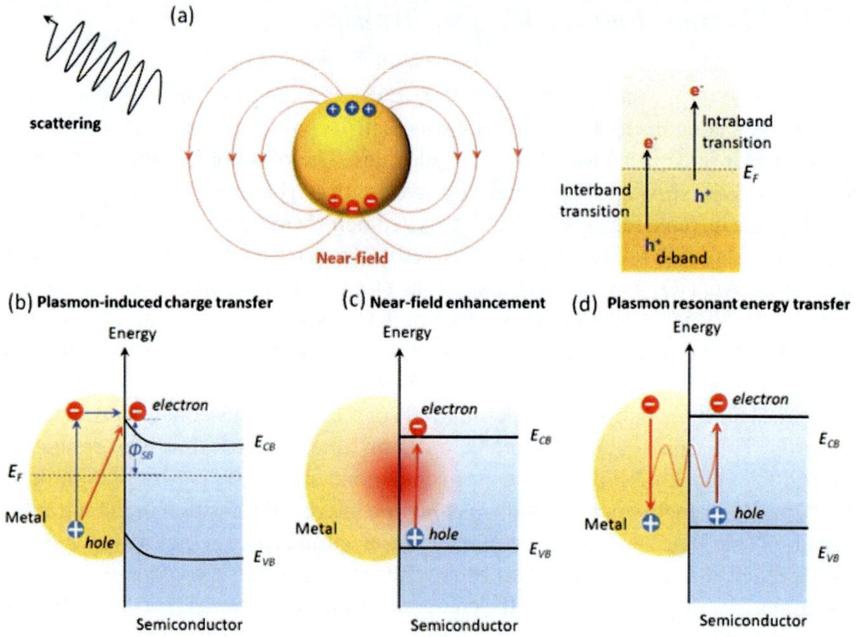

Fig. 3.1 Schematic illustrations of the mechanism of the plasmonic enhancement effect. **a** LSPR of metal NPs. **b–d** present the plasmon-induced charge transfer, near-field enhancement and plasmon resonance energy transfer mechanisms, respectively

ers is short, the reported efficiencies for devices based on plasmon-induced charge separation concepts are still low. One approach towards solving this issue is to create a direct metal-to-semiconductor interfacial charge transfer transition that can be directly excited to promote an electron from the metal into the conduction band of the semiconductor, as indicated by the red arrow in Fig. 3.1b; this process is also called plasmon-induced direct charge transfer. It is an extension of the chemical interface damping that is used to explain the adsorbate-induced resonance band broadening of plasmonic metal NPs. A key feature to achieve this direct charge transfer is the strong interdomain coupling and mixing of the metal and semiconductor energy levels. This proposed electron transfer mechanism involves a new plasmon decay pathway: the direct excitation of an electron in the semiconductor and a hole in the plasmonic metal. Using cadmium selenide nanorod-capped Au rods, the plasmon was strongly damped by cadmium selenide through interfacial electron transfer. The quantum efficiency was shown to be higher than 24% and independent of the excitation photon energy [40].

3.2.2 Near-Field Enhancement

Upon irradiation, the coherent free electrons of plasmonic metal NPs oscillate to produce an ultrahigh local electromagnetic field. The excited plasmonic metal NP behaves as a concentrator of the local near field, and the intensity of the local electromagnetic field is dramatically boosted near the interface of the metal NP and the semiconductor. If the resonant energy of the LSPR overlaps with the absorption band of the semiconductor, the interband transition of the semiconductor will be largely enhanced. Electromagnetic simulations have shown that the electromagnetic field intensity of the local near field can reach as much as a thousand times that of the incident field [41]. As a consequence, the rate of charge carrier generation is increased a thousand times. Thus, an increased amount of photoinduced charge is generated locally due to the local field enhancement of the plasmonic NPs, as shown in Fig. 3.1c. This enhancement mechanism occurs on metal NP-loaded small-bandgap semiconductors, such as Fe_2O_3 and CdSe [42, 43]. In the case of TiO_2, this enhancement mechanism relies on the presence of defect states located in the bandgap of TiO_2 to enable light absorption via the electronic transition from the defect states to the conduction band. Because the intensity of the local near field decays exponentially as a function of the distance away from the plasmonic metal NP, this mechanism takes place only within a certain distance around the plasmonic metal NP.

3.2.3 Plasmon Resonance Energy Transfer

Plasmon resonance energy transfer (PRET) refers to a nonradiative mechanism of LSPR-induced charge separation via a dipole-dipole energy transfer process [26]. It is analogous to Förster resonance energy transfer, but the LSPR dipole replaces the fluorescent system. In the PRET process, the dipole moment of the electron-hole pair of the semiconductor is coupled with the plasmonic dipole, and the electron-hole pairs are nonradiatively excited in the semiconductor via relaxation of the plasmon dipole (see Fig. 3.1d). PRET can generate electron-hole pairs in the semiconductor using photon energies both above and below the semiconductor bandgap because of the nonradiative coupling between the plasmon dipole and the optically inefficient states at the band edge of the semiconductor. There are two main advantages of charge carrier generation under PRET. First, it is not limited by the electronic band structure alignment of the semiconductor and the plasmonic metal NP and charge equilibration. It might be helpful to understand it by comparing it to the plasmon-induced charge transfer. For plasmon-induced charge transfer on an n-type semiconductor, the Fermi level of the metal NP should align at a lower energy than the conduction band of the semiconductor to form a Schottky barrier for efficient charge separation. For PRET, because the electron transition occurs in the semiconductor, there is no restriction on the band alignment between the metal NP and semiconductor. Second, it can use the energies both above and below the semiconductor bandgap, overcoming the absorption limitation of the semiconductor. Therefore, it offers flexibility for designing solar energy conversion materials and devices.

3.3 Plasmonic Water Splitting

Photocatalytic and PEC water splitting are two common types of plasmonic water splitting systems. In the plasmonic photocatalytic water splitting system, plasmonic metal NP-decorated semiconducting photocatalytic powders are simply dispersed in water, and hydrogen and oxygen are then readily produced upon illumination [44, 45]. Although this system is simple and suitable for large-scale solar water splitting applications, the system experiences some problems. The most challenging one is the difficulty of separating the evolved H_2 from the gas mixture to avoid the back reaction between H_2 and O_2. In the plasmonic PEC system, plasmonic metal NPs are generally fabricated on the surface of a semiconductor as a photoelectrode to implement solar energy conversion with suitable light energy [31, 44, 46–49]. Since H_2 evolution and O_2 evolution take place in two separate chambers, the back reaction between H_2 and O_2 is prevented. In this section, we introduce the plasmonic water splitting system based on the PEC water splitting system.

3.3.1 Plasmonic Water Oxidation

Regarding water splitting reactions, water oxidation is a bottleneck reaction for photochemical water splitting because it is a thermodynamically unfavorable and kinetically difficult four-electron chemical reaction process. To implement an efficient plasmonic water splitting system, much attention should be focused on the plasmonic water oxidation half-reaction.

With the development of nanofabrication technologies, bottom-up and top-down methods have been rapidly developed to fabricate plasmonic metal nanostructures on the surface of semiconductor NPs or semiconductor bulk materials [50–52]. Based on sophisticated electron beam lithography (EBL) and lift-off technologies, the author's group has developed a well-designed Au nanorod array on n-type TiO_2 as a plasmonic photoanode to study the plasmonic water oxidation reaction [31]. Fig. 3.2a shows scanning electron microscopy (SEM) images of Au nanorods with a size of 100×200 nm and height of 30 nm on TiO_2. The Au nanorod array was precisely controlled and had intimate contact with the crystal surface of TiO_2. To characterize the PEC properties, a typical three-electrode electrochemical measurement system was employed to measure the incident photo-to-current conversion efficiency (IPCE) of this plasmonic photoelectrode. In the measurement system, a Pt wire was used as the counter electrode, a saturated calomel electrode was used as the reference electrode, and Au nanorod-loaded TiO_2 was used as the working electrode. Because only a 0.1

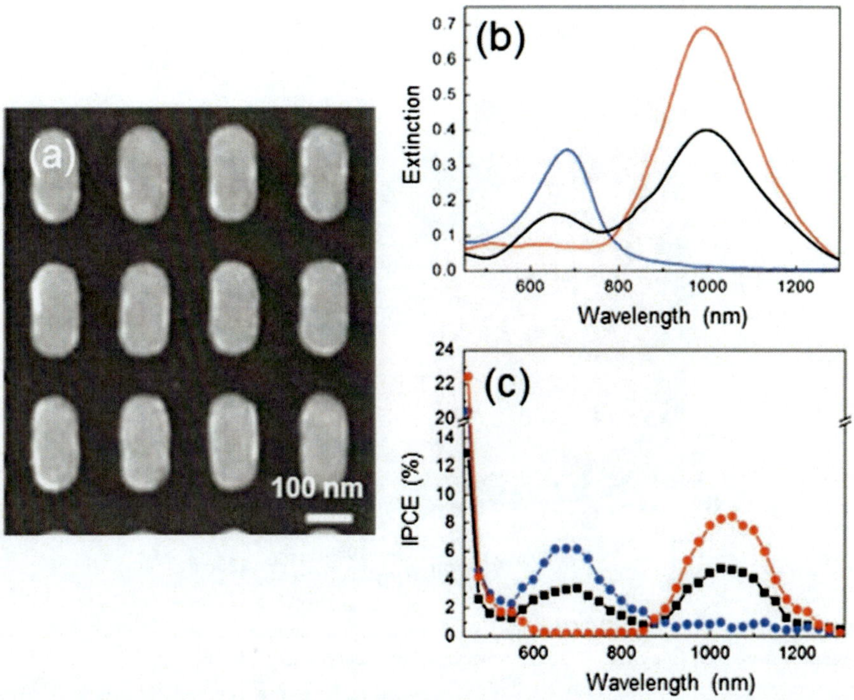

Fig. 3.2 **a** SEM image of Au nanorods on a TiO_2 surface. **b** Extinction spectra and **c** IPCE action spectra of Au nanorods/TiO_2 under polarized and nonpolarized light irradiation. The nonpolarization, horizontal polarization, and vertical polarization curves are shown in black, blue, and red, respectively. Adapted from [31] Copyright © 2010, American Chemical Society

M $KClO_4$ solution was used as the supporting electrolyte, the only possible PEC reaction in this reaction environment is the water oxidation reaction on this photoanode.

Figure 3.2b shows the extinction spectra of the Au nanorods measured in the electrolyte solution under nonpolarized and polarized light irradiation from visible to near-infrared (NIR). Two distinct LSPR bands were clearly observed that corresponded to the transverse mode (blue) and longitudinal mode (red). Figure 3.2c shows the action spectra of the incident photon-to-current conversion efficiency (IPCE) under nonpolarized and polarized light irradiation. The shapes and peak wavelengths of the IPCE action spectra closely match those of the plasmon resonance bands. According to the measurement of the IPCE action spectrum under linearly polarized conditions, the IPCE values of the photocurrent were up to 6.3 and 8.4% near the LSPR bands. These observations suggested that electron injection from the Au nanorods to TiO_2 originated from the LSPR.

By employing this Au nanorod-loaded TiO_2 photoelectrode, we further quantitatively explored the oxygen gas and hydrogen peroxide as the products of water oxidation that evolved from these plasmonic photoelectrodes [46]. Water containing the ^{18}O isotope was used, and the amount of $^{34}O_2$ was determined using gas

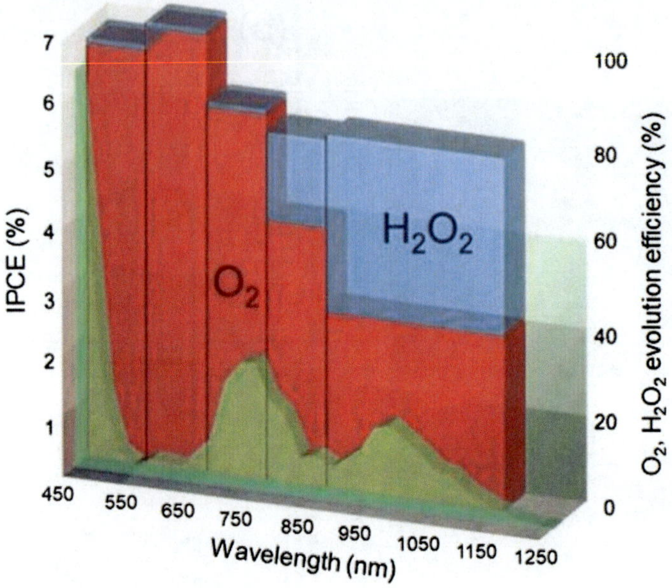

Fig. 3.3 IPCE action spectrum and the gas evolution efficiency under visible to NIR irradiation. Adapted from [46] Copyright © 2012, American Chemical Society

chromatography-mass spectrometry (GC-MS) to quantitatively determine the oxygen evolution. Hydrogen peroxide was also quantitatively determined by absorptiometry using oxo[5,10,15,20-tetra(4-pyridyl)porphyrinato]titanium(IV) as an indicator for hydrogen peroxide detection. Figure 3.3 presents the rate of the evolved O_2 and H_2O_2 as a function of the irradiation wavelength in the visible to near-infrared range. Four-electron water oxidation occurred in the shorter-wavelength region, while two-electron oxidation of water molecules was observed in the longer-wavelength region because two-electron oxidation of water for H_2O_2 evolution is kinetically easy to be induced than four-electron oxidation for O_2 evolution. Interestingly, water oxidation occurred even under NIR irradiation with photon energy close to the lowest energy limit for water splitting (1.23 eV). It indicates that plasmonic metal NP may act as an efficient element to lower the overpotential of the water oxidation reaction, possibly caused by the concentrated electron/hole charge carriers excited by the ultrahigh local field near the plasmonic metal NP.

According to the charge transfer between Au NPs and TiO_2 demonstrated by ultrafast transient absorption measurements, [25] a possible plasmon-induced charge transfer mechanism for efficient water oxidation reactions was proposed. As illustrated in the energy diagram in Fig. 3.4, an excited electron is transferred to the conduction band of TiO_2 immediately after the interband and/or intraband transition of Au NPs induced by the ultrahigh local electromagnetic field. Because the lifetime of plasmon-induced charge carriers is much shorter than the time scale of chemical reactions, the direct participation in chemical reactions of these charge carriers is unfavorable.

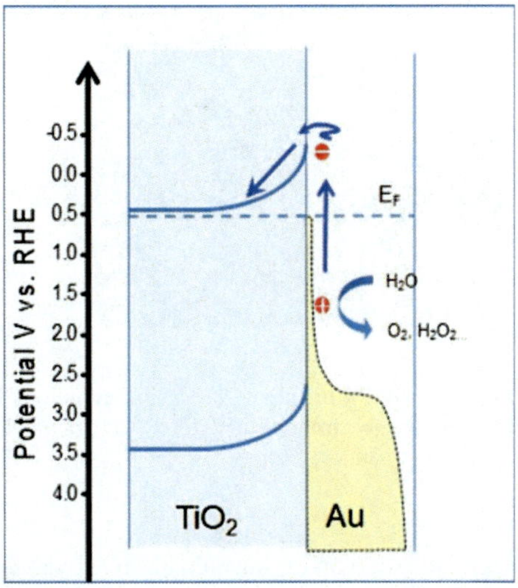

Fig. 3.4 Possible mechanism of plasmon-induced charge separation at the Au/TiO_2 interface

Instead, the plasmon-induced hole might be trapped in the surface states of TiO_2 near the three-phase boundary of Au/TiO_2/water and subsequently participate in the water oxidation reaction. On the basis of this mechanism, the plasmon-enhanced ultrahigh electromagnetic near field would highly promote water oxidation, resulting in the evolution of O_2 and H_2O_2 even at low-energy wavelengths in the NIR region.

To further understand the mechanism of plasmonic chemical reactions, it is important to study the plasmon-induced near field in the spatial and time domains. Photoemission electron microscopy (PEEM) serves as a powerful tool to explore the spatial distribution of the plasmonic near field as well as the charge carrier transport pathway [53–55]. When combining the PEEM with pump-probe technology, it is also possible to investigate the lifetime of the plasmon resonance and further explore the energy distribution of the energetic charge carrier. Future studies on this subject may deepen our understanding of the mechanism of plasmonic chemical reactions.

The fabrication efficiency of precisely designed metal nanostructures is generally low. A high-throughput method to fabricate plasmonic photoelectrodes is generally needed for practical applications. Park et al. reported the fabrication of Au NP (5–9 nm)-decorated ZnO-TiO_2 core-shell nanowires by a photodeposition method for PEC waster splitting with a maximum PEC water splitting efficiency of 0.7% at +0.8 V vs RHE was demonstrated [56]. Li et al. reported a largely enhanced PEC performance on Au NP-decorated three-dimensional TiO_2 nanorod architectures fabricated by surface-reaction-limited pulsed chemical vapor deposition [57]. The Au NPs were fabricated by both sputtering and chemical reduction methods. To achieve high-throughput metal NP fabrication, the author's group also developed a simple Au NP fabrication method by annealing Au thin films deposited on semiconductor surfaces to investigate plasmonic water oxidation [58]. Single-crystal TiO_2 (rutile, 0.05 wt% Nb doped) with a (110) surface orientation was used in the study. A 3-nm Au film

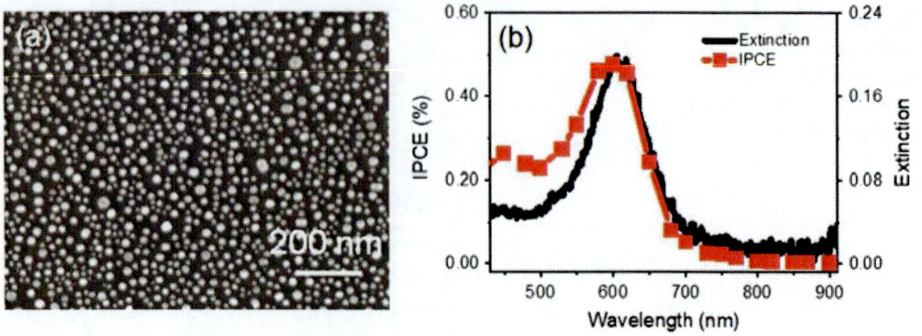

Fig. 3.5 **a** SEM image of Au NP-loaded onto a TiO_2 single crystal. **b** Extinction spectrum and IPCE action spectrum of the Au NP/TiO_2 photoelectrode. Adapted from [58] Copyright © 2012, American Chemical Society

was deposited on the surface of TiO_2 by sputtering and annealed at different temperatures in a nitrogen atmosphere to form Au NPs on the TiO_2 surface, as the SEM image shown in Fig. 3.5a. Under the PEC measurement using a three-electrode system similar to that mentioned above, the IPCE action spectrum closely matched the extinction spectrum, with a maximum value of 0.5%, as presented in Fig. 3.5b.

As the possible mechanism discussed above, the plasmon-enhanced photocurrent generation may result from the plasmon-induced electron transfer from Au NPs to the conduction band of TiO_2. A hole with a high oxidation ability was trapped at the TiO_2 surface states near the Au NPs/TiO_2 interface, which has a positive potential for water oxidation. To confirm the oxygen evolution, the $^{34}O_2$ was determined by GC-MS using ^{18}O water. The intensity of the peak at m/z 34 increased with irradiation time, as shown in Fig. 3.6a. The linear relationship between the number of evolved oxygen molecules and the number of photogenerated electrons was clearly observed in Fig. 3.6b. The yield of evolved O_2 as a function of the observed photocurrent was estimated to be 91.6%, indicating that the O_2 evolution via four-electron oxidation of water molecule proceeded stoichiometrically.

One advantage of this Au NP fabrication method is that the interfacial structure can be controlled via the annealing temperature. Under high-temperature annealing, an intimate interfacial contact is formed, which is generally requested for efficient charge transfer. We will discuss this part in more detail in Sect. 3.2.

The crystal phase bulk semiconductor generally produces a high energy conversion efficiency because of its low number of structural defects. However, the cost of bulk crystal semiconductor substrates is commonly a key criterion, and their high energy consumption during fabrication is unacceptable for widespread applications. To overcome these disadvantages, one alternative strategy is to use a thin film semiconductor instead of the bulk crystal. The abundant structural defects in the thin film structure decrease the energy conversion efficiency relative to that of the crystal phase semiconductor, and the cost efficiency of the thin film-based energy conversion system is increased rapidly. Thus, thin film-based solar energy conversion systems are attracting increasing research interest, especially in photovoltaics [59–61].

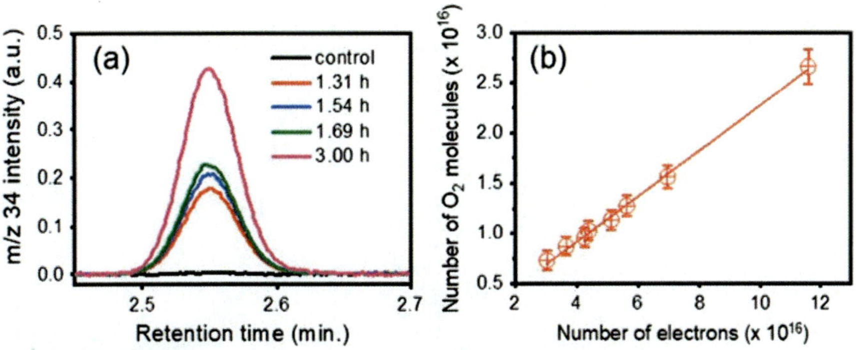

Fig. 3.6 **a** Chromatogram of the $^{34}O_2$ detected by the GC-MS. **b** Evolved oxygen molecule number as a function of the electron number. Adapted from [58] Copyright © 2012, American Chemical Society

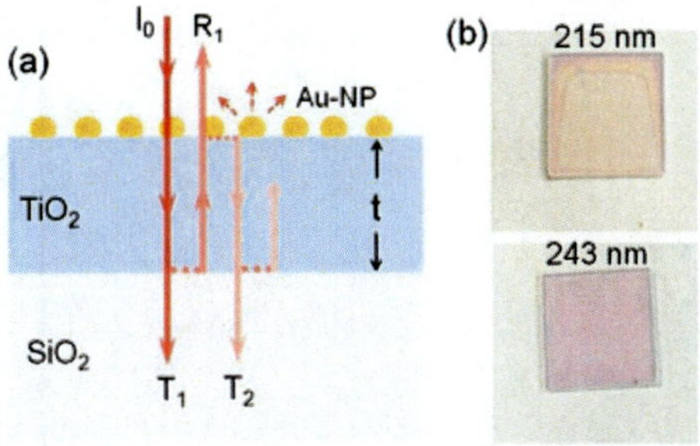

Fig. 3.7 **a** Schematic illustration of the Fabry-Perot interference in the TiO_2 film. **b** Photographs of the TiO_2 thin film structures with film thickness of 215 and 243 nm. Adapted from [62] Copyright © 2013, American Chemical Society

Based on the concept of thin film plasmonic photoelectrodes, the authors' group fabricated a TiO_2 thin film structure by atomic layer deposition (ALD) instead of a costly single-crystal TiO_2 for plasmonic water oxidation [62]. TiO_2 thin films with thicknesses of less than 300 nm were deposited onto a SiO_2 substrate using a commercial hot-wall flow-type ALD reactor involving alternating exposure of $TiCl_4$ and water vapor at a temperature of 300 °C with N_2 as a precursor carrier and purge gas. The TiO_2 film exhibited a characteristic color caused by the Fabry-Perot interference in the high-refractive-index TiO_2 thin film, as shown schematically in Fig. 3.7a. Color photographs of different thicknesses of TiO_2 are presented in Fig. 3.7b. The XRD measurements indicated that the crystallographic phase of the TiO_2 thin films was anatase-type TiO_2, which shows better photocatalytic activity than the rutile phase. In addition, the LSPR of Au NPs on TiO_2 thin films can be strongly enhanced by Fabry-Perot interference when the cavity resonance wavelength overlaps with the plasmon band.

To demonstrate the interference effect on the LSPR as well as the IPCE, TiO_2 thin films with resonance near and far from the LSPR resonance wavelength were fabricated. As shown in Fig. 3.8a, the absorption spectrum of the 215-nm TiO_2 thin film shows an obvious Au NP plasmon resonance band at approximately 600 nm due

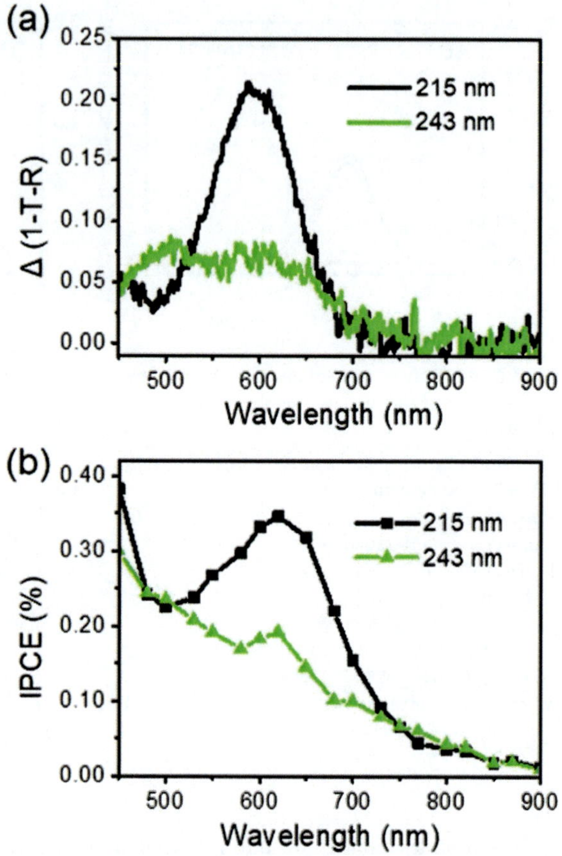

Fig. 3.8 **a** Absorption spectra and **b** IPCE spectra of Au NP/TiO_2 photoelectrode. Adapted from [62] Copyright © 2013, American Chemical Society

to overlap between the cavity resonance and the LSPR. In contrast, for the 243-nm TiO_2 film, an unobvious absorption band was observed due to the negligible overlap between the cavity resonance and the LSPR. The IPCE action spectra of Au NPs/TiO_2 with thicknesses of 215 and 243 nm are shown in Fig. 3.8b. The IPCE spectrum of the 215-nm TiO_2 film reveals an obvious enhancement band at around 620 nm, while the IPCE of the 243-nm TiO_2 film shows only a small hump. Namely, the interference in TiO_2 can modulate the plasmon resonance as well as the PEC water oxidation reactions.

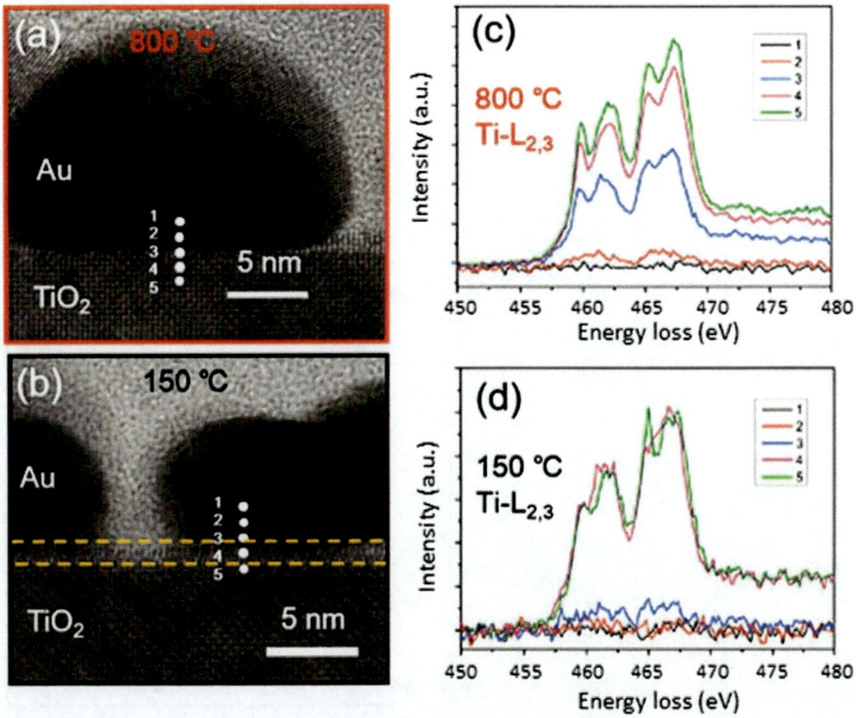

Fig. 3.9 **a, b** HR-TEM images and **c, d** corresponding Ti-$L_{2,3}$ EELS spectra of Au NP/TiO_2 under low- and high-temperature annealing. Adapted from [58] Copyright © 2012, American Chemical Society

3.3.2 Interfacial Structure Effects on Plasmonic Water Oxidation

On the plasmonic photoelectrode, plasmon-induced electron transfer is believed to be a possible mechanism for plasmonic water splitting. For efficient charge transfer at the interface of the metal/semiconductor, an intimate interfacial contact is thus required. The interfacial structure thereby plays an essential role in the plasmon-induced charge transfer at the interface of the metal/semiconductor. In our group, the interfacial structure effect on plasmon-induced charge transfer was also studied for the future design of an efficient PEC water splitting system [58, 63].

The effect of interfacial defects at the Au NP/TiO_2 heterojunction on the plasmon-induced photocurrent was investigated in detail [58]. Au NPs were fabricated on TiO_2 single crystals by 3-nm Au film annealing at 150 and 800 °C. Figure 3.9a, b shows the cross-section images of Au NPs/TiO_2 upon 800 and 150 °C annealing obtained by high-resolution transmission electron microscopy (HR-TEM). Upon 800°C annealing, Au NPs were in tight contact with the crystalline surface of TiO_2, while upon

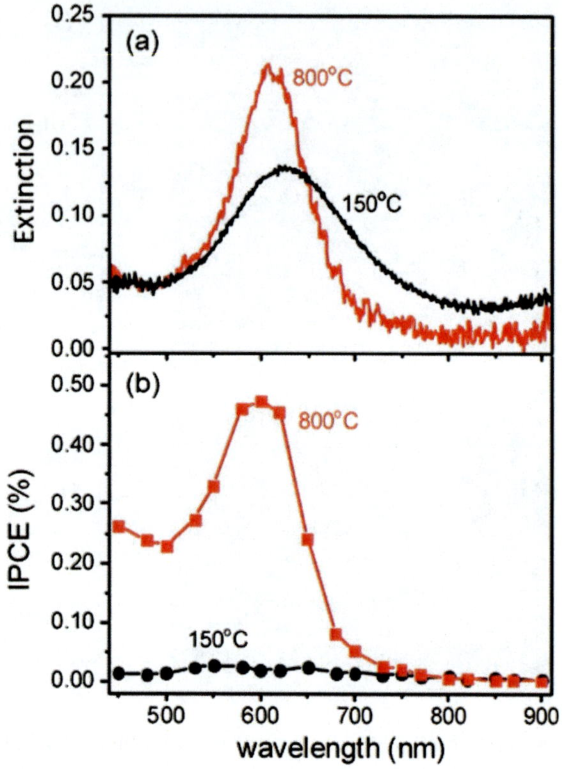

Fig. 3.10 **a** Extinction spectra and **b** IPCE spectra upon low- and high-temperature annealing. Adapted from [58] Copyright © 2012, American Chemical Society

150°C annealing, an interfacial impurity on the order of a few atomic layers was observed, as the yellow dashed line marked. Electron energy loss spectroscopy (EELS) was performed to further confirm that the interfacial layer formed upon 150°C annealing is a reduced TiO_2 defect layer, as depicted in Fig. 3.9c, d. Because the hole-capture ability of the TiO_2 defect layer is weak, the PEC performance of Au NP/TiO_2 upon 150°C annealing dramatically decreased, as determined by IPCE measurements. Figure 3.10a, b show the extinction spectra and IPCE action spectra of Au NPs/TiO_2 upon 150 and 800 °C annealing, respectively. Although the extinction values at the resonance wavelength are slightly different, the IPCE value at the LSPR wavelength around 600 nm dramatically increased upon 800 °C annealing, which also shows an obvious IPCE band. Notably, there was no obvious IPCE band for the Au NPs/TiO_2 annealed at 150°C due to the low hole-capture ability of the TiO_2 defect layer at the interface. Conclusively, intimate interfacial contact between metal NPs and semiconductors is in high demand for efficient plasmon-induced charge separation.

Because the interfacial structure is ultrasensitive to the surface termination layer of the semiconductor, it is also possible to modulate the interfacial structure by depositing metal NPs on the surface of the crystal semiconductor with different surface orientations. Since STO is a model of perovskite-type oxides having the general formula

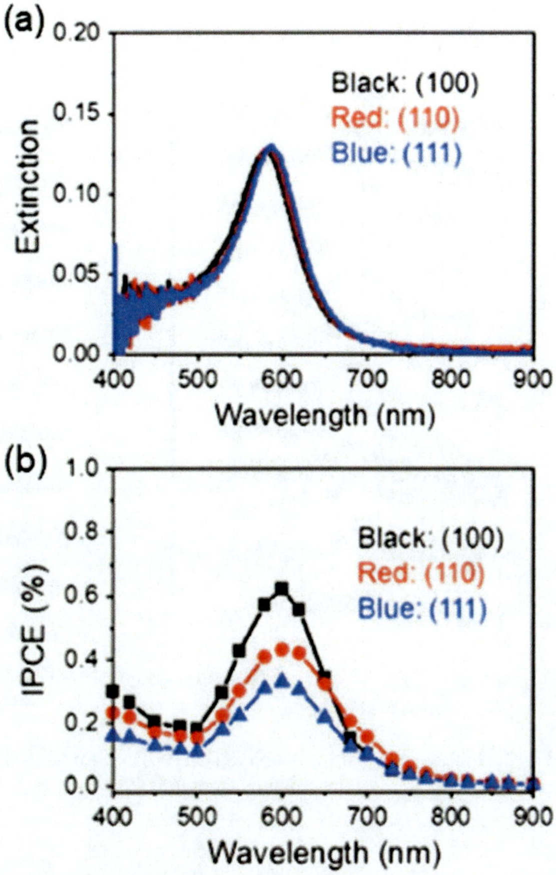

Fig. 3.11 **a** Extinction spectra and **b** IPCE action spectra of Au NPs loaded on STO with three kinds of oriented surfaces. Adapted from [63] Copyright © 2020, American Chemical Society

ABO_3, our group also selected STO to investigate the interfacial structure effect on plasmon-induced photochemical reactions [63]. When Au NPs settled on the surface of STO, the interfacial surface states could be modulated via the surface termination layer of STO. Consequently, the PEC activities of Au NP/STO can be controlled. In this study, Au NPs were fabricated on STO with (100), (110), and (111) surface orientations. Because the particle distribution was almost the same in all Au NP/STO samples, the extinction spectra were almost identical, as presented in Fig. 3.11a. However, their PEC activities are remarkably different, as shown in Fig. 3.11b. The IPCE of Au NP/STO (100) shows a remarkably higher IPCE than the other two orientation surfaces. This experiment showed that surface termination layer engineering is a feasible way to manipulate plasmon-induced chemical reactions.

To further understand the interfacial effect on the plasmon-induced chemical reaction, the intermediate of water oxidation on Au NP/STO with surface orientations of (100) and (110) was monitored during the plasmon-induced water oxidation reaction using in situ electrochemical SERS. Figure 3.12 shows the potential-dependent SERS

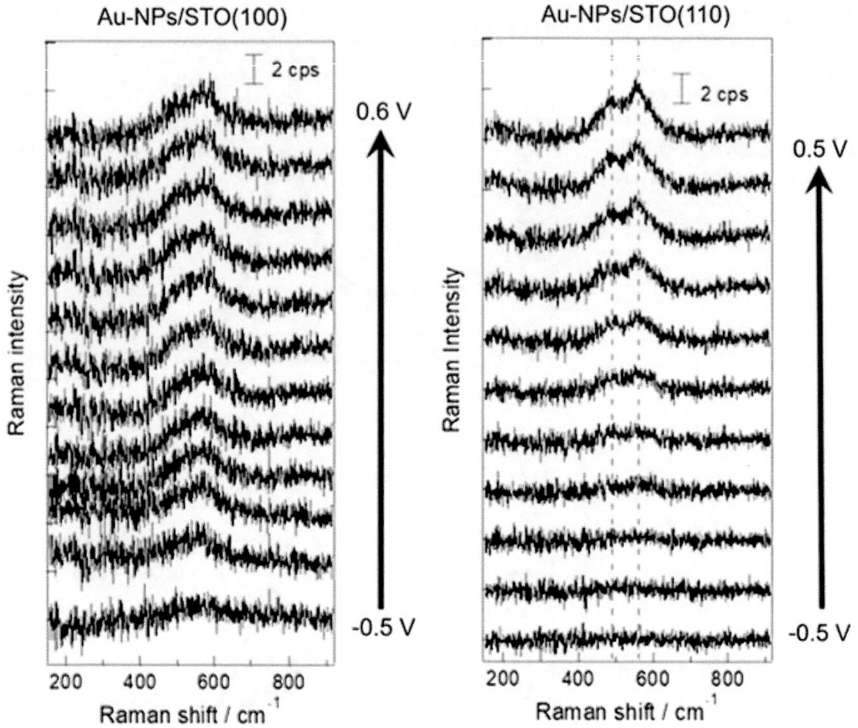

Fig. 3.12 Raman signals of Au NP/STO (100) and Au NP/STO (110) measured in an in situ electrochemical SERS system. Adapted from [63] Copyright © 2020, American Chemical Society

spectra of Au NPs/STO (100) and (110). The Au-O stretching modes were monitored for the high energy AuOOH intermediate of water oxidation. The formation of Au-O bonds on the Au surface is observed at approximately 0.21 V vs. RHE on the (100) facet, while it starts at approximately 0.61 V vs. RHE on the (110) facet. The negative onset potential of the SERS signal of Au-O vibrations demonstrated that Au NP/STO (100) underwent a much more effective water oxidation at a lower overpotential against Au NP/STO (110). Based on the IPCE and in situ SERS measurements, the higher plasmon-induced water oxidation ability on Au NP/STO (100) was attributed to the positive surface states and efficient oxidation activity at the interface between Au NP and STO (100).

3.3.3 Self-assisted Plasmonic Water Splitting

In most PEC water splitting systems, an externally applied potential is needed to overcome the overpotentials of the hydrogen and oxygen evolution reactions. The

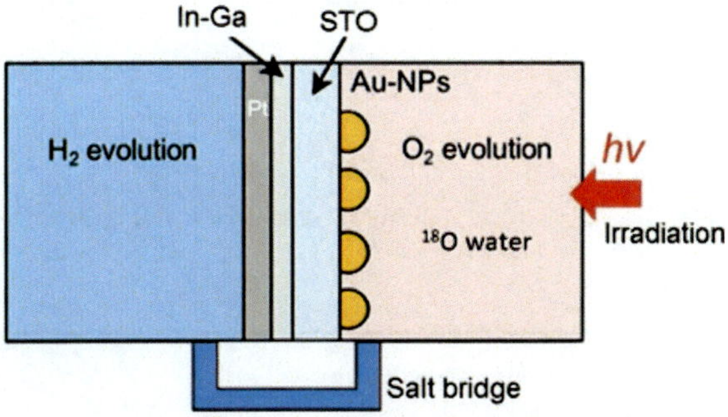

Fig. 3.13 Schematic of the photochemical reaction cell using two sides of STO. Adapted from [64] Copyright © 2014, WILEY-CH Verlag GmbH & Co. KGaA

request of the externally applied potential makes the PEC water splitting systems more challenging in addition to improving the chemical reaction efficiency. Thus, self-assisted plasmonic water splitting without external bias has attracted attention. Moskovits and coworkers fabricated a Au nanorod array capped with TiO_2 and cocatalysts using template-lithographic technologies for autonomous complete water splitting [32]. The solar-to-hydrogen efficiency was reported to be 0.1% for this system. Our group utilized the front and back sides of the same strontium titanate ($SrTiO_3$, STO) single-crystal substrate as an anode and a cathode for oxygen and hydrogen evolution, respectively. [64] A schematic illustration of the plasmonic water splitting system used in the study is shown in Fig. 3.13.

On the anode side, Au NPs were fabricated on an STO single-crystal substrate by Au thin film annealing methods; on the cathode side, a platinum plate was pasted as a hydrogen evolution catalyst. Under this construction, the anodic and cathodic side chambers were separated, and the evolved gases were generated separately in the respective chambers. To maintain the charge balance between the anode and cathode chambers, these two chambers were joined with a salt bridge. The evolved H_2 and O_2 gas was quantitatively identified by GC and GC-MS according to the methods described in the previous sections. Instead of the electrical bias in the traditional PEC system, in this study, a pH bias between the anode and cathode was used to drive the water splitting reactions. The acid and alkaline electrolytes were employed on the anode and cathode sides to build up a chemical bias. When the photoanode was illuminated with visible light from 450 nm to 850 nm, the amount of the evolved hydrogen and oxygen gases increased linearly with the irradiation time, as shown in Fig. 3.14a. The evolved gas volume was determined to be 2:1, which indicates complete water splitting reactions. The plasmonic effect on the water splitting reactions was further confirmed by measuring the action spectrum of the gas evolution. The H_2 evolution action spectrum, as shown in Fig. 3.14b, closely matches the plasmon resonance spectrum.

A possible plasmon-induced charge transfer mechanism for this self-assisted plasmonic water splitting is illustrated in Fig. 3.15. Plasmon-induced charge separation is induced at the interface of Au NP/STO, followed by electron transfer from Au NP to the conduction band of STO. Under the established chemical bias, the injected elec-

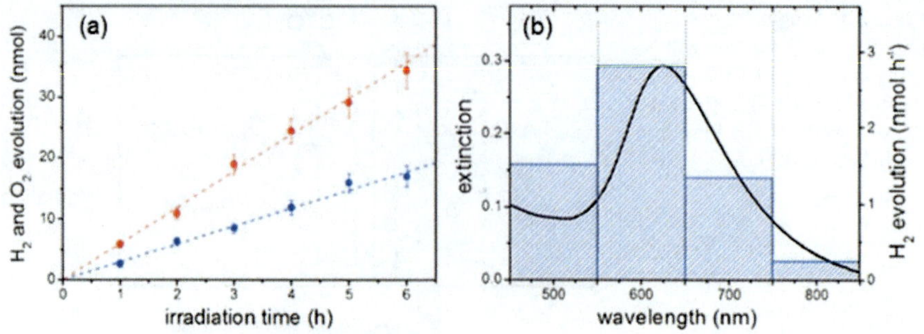

Fig. 3.14 **a** Hydrogen and oxygen evolution on Au NP/STO under visible light irradiation. **b** Hydrogen action spectrum versus the extinction spectrum. Reprinted from [64] Copyright © 2014, WILEY-CH Verlag GmbH & Co. KGaA

trons are driven to the platinum cathode side for hydrogen evolution. The holes oxidize hydroxyl ions or water molecules to induce oxygen evolution. As a result, hydrogen gas and oxygen gas can be separately evolved from two sides of the STO substrate, which is beneficial for the utilization of H_2 as an energy carrier gas. In this proposed mechanism, one can see that the chemical bias is also very important for driving the photochemical reactions. Further experiments were carried out with various chemical bias combinations between the anode and cathode sides. The minimum chemical bias that can sustain the water splitting reaction in this system is 230 mV. Note that this is a small bias that needs to drive a complete water splitting reaction. Generally, the overpotential for the water oxidation reaction is several hundred mV because it is a kinetically difficult reaction process. The small bias for the water splitting reactions here might benefit from efficient water oxidation under plasmon resonance.

3.3.4 Plasmonic Water Splitting on Three-Dimensional Structures

From the viewpoint of the active chemical reaction site, the three-dimensional structure is beneficial for enhancing the photochemical energy conversion efficiency. Zhang et al. employed a TiO_2-based photonic crystal three-dimensional structure for plasmonic PEC water splitting [33]. The design of the three-dimensional structure significantly increased the LSPR of Au NPs and consequently boosted the electron injection from Au NPs to TiO_2 and thus enhanced the PEC water splitting performance under visible light irradiation. Yen et al. used a three-dimensional structure constructed by Au NPs on a TiO_2 hierarchical network to enhance photocurrent generation [65].

A three-dimensional TiO_2 nanotunnel (TNT) structure that can enhance light absorption and the reaction area was also proposed by the author's group [66] TNTs

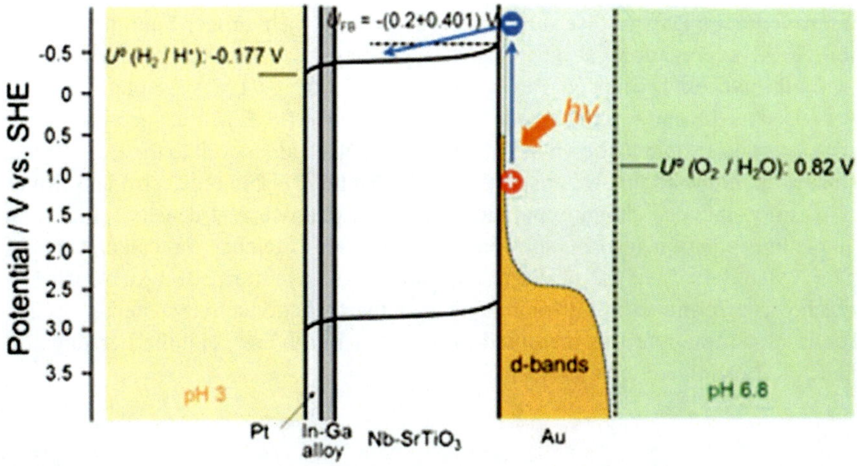

Fig. 3.15 Illustration of the mechanism of the water splitting reaction on Au NP/STO. Reprinted from [64] Copyright © 2014, WILEY-CH Verlag GmbH & Co. KGaA

are promising because of their unique high electron mobility, large specific surface area, and high mechanical strength, especially rapid electron transport through the walls of self-assembled nanotunnels. A three-dimensional plasmonic photoanode consisting of Au NP-decorated TNTs (Au NPs/TNTs) was developed for PEC water splitting. To fabricate the TNT structure, a Ti plate was chemically polished in an etchant and anodized in 0.38 wt% NH_4F ethylene glycol solution under an anodizing voltage of 60 V. Figure 3.16a shows the side-view SEM images of the TNTs constructed with 102 nm diameter nanoholes and a length of 1 μm. Au NPs were deposited by a chemical reduction method to achieve a high surface coverage of Au NPs.

A cross-sectional scanning transmission electron microscopy (STEM) image of the Au NPs/TNTs prepared from the $HAu(OH)_4$ precursor is shown in Fig. 3.16b. The Au NPs were well dispersed because the hydroxyl group of $HAu(OH)_4$ strongly

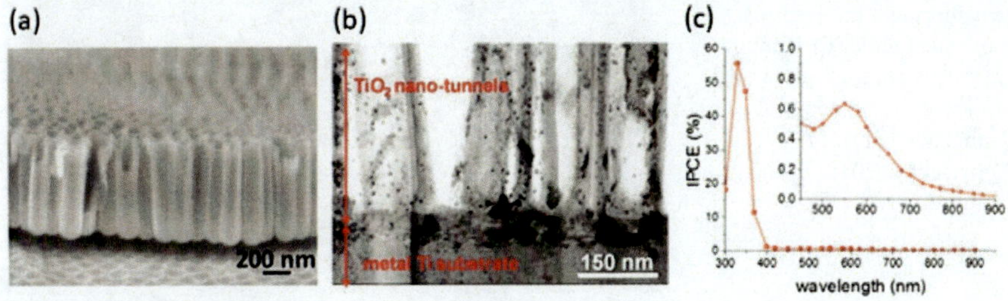

Fig. 3.16 **a** Side view of the three-dimensional TNT structure. **b** Cross-section of Au NP-decorated TNTs. **c** IPCE action spectrum of the Au NP/TNT photoelectrode. Adapted from [66] Copyright © 2017, Royal Society of Chemistry

adsorbs onto the TiO_2 surface and prevents desorption via hydrogen bonding. In addition, Au NP aggregation was prevented during their growth due to the Coulomb repulsion of the hydroxyl group. For PEC measurements, Au NPs/TNTs and platinum coils were used as the anode and cathode, respectively. The IPCE action spectrum of the two-electrode system is shown in Fig. 3.16c and corresponds well to the LSPR band. This result indicated that water splitting in the Au NP/TNT photoelectrode is driven by plasmon-induced charge separation. From the photocurrent density under solar simulator irradiation, the PEC solar energy conversion efficiency was calculated to be 0.1%. Quantitative analysis of H_2 gas evolution was also performed on Au NPs/TNTs, which was determined by GC mounted with a thermal conductivity detector (TCD). The H_2 evolution rate was measured to be 1.05 μmol/h/cm^2, and the Faraday efficiency was estimated to be 73.9%.

3.3.5 *Water Splitting Under Strong Coupling Conditions*

The general plasmonic photoelectrode for PEC water splitting consists of a single layer of metal NPs on the semiconductor surface. The absorption of a single layer of metal NPs is usually insufficient, especially in the case of small metal NPs (< 50 nm). Constructing a plasmonic photoelectrode with high absorption over a broad energy region is essential to improve the solar energy conversion efficiency. In addition, water oxidation, a bottleneck reaction of water splitting, is thermodynamically unfavorable because it is a kinetically difficult four-electron chemical reaction process. It is also important to enhance the quantum efficiency of water oxidation for efficient water splitting.

Aiming to solve these two challenges, we developed an Au NP-loaded TiO_2 thin film photoelectrode under strong coupling conditions [49]. The photoelectrode consists of a Au NP/TiO_2/Au film (ATA) structure with partially inlaid Au NPs, as shown

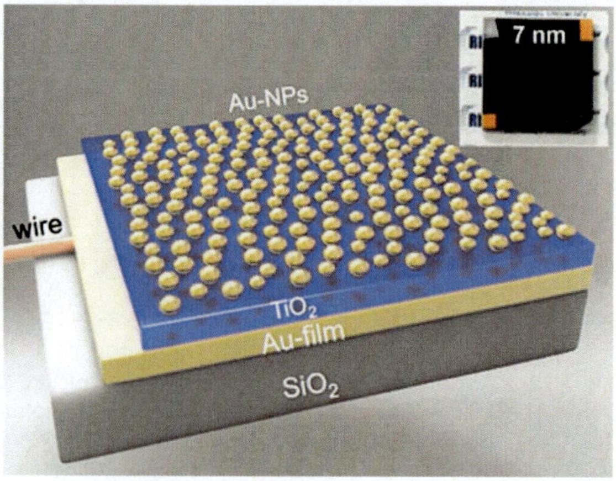

Fig. 3.17 Schematic illustration of the ATA structures and the electron movement under irradiation. Insert is a photograph of the ATA with inlaid depth of 7 nm. Adapted from [49] Copyright © 2018, Springer Nature

in the schematic illustration of Fig. 3.17a. The strong light absorption is caused by the strong interaction between the LSPR of Au NPs and the cavity resonance. Generally, when an absorber settles on top of a cavity, the interaction between them is weak due to the small overlapping mode volume. To increase the interaction, partially inlaid Au NPs into TiO_2 were engineered by post deposition of a TiO_2 layer using ALD. Insert in Fig. 3.17a shows photographs of the ATA samples with inlaid depths of 7 nm. The ATA structures appear black, showing strong light absorption. As discussed in Sect. 3.1, because TiO_2 has a large refractive index as well as the reflection phase changes at the TiO_2/Au film interface, a TiO_2/Au film can support cavity mode with a TiO_2 thickness of only several tens of nanometers. When the LSPR of Au NPs overlaps the cavity resonance, this system may step into a strong coupling regime.

Strong coupling can be observed when two resonators interact with each other strongly enough that the eigenfrequencies of the coupled system shift away from the degenerate eigenfrequencies of the uncoupled resonators. Under such strong coupling conditions, substantial modification of characteristic properties beyond the chemical reactivity, such as the electrical conductivity, [67] energy transfer rate, [68] and reverse intersystem crossing, [69] are observed. A characteristic feature of strong coupling is energy level splitting, a property that can be well understood from classical theory. To understand the strong coupling effect, ATA structures (7-nm inlaid) with different TiO_2 thicknesses (19 to 44 nm) were constructed. To analyze the splitting energy, a dispersion curve was obtained by plotting the peak energy of the absorption spectra as a function of the cavity resonant energy, as shown in Fig. 3.18a. From the dispersion curve, the energy splitting was calculated to be as large as 320 meV using a coupled harmonic oscillator model. Accordingly, the energy diagram of this strongly coupled system is proposed, as shown in Fig. 3.18b. When the cavity resonant mode in TiO_2 overlaps the LSPR of the Au NPs, hybrid modes with upper and lower branches separated by energy splitting are formed.

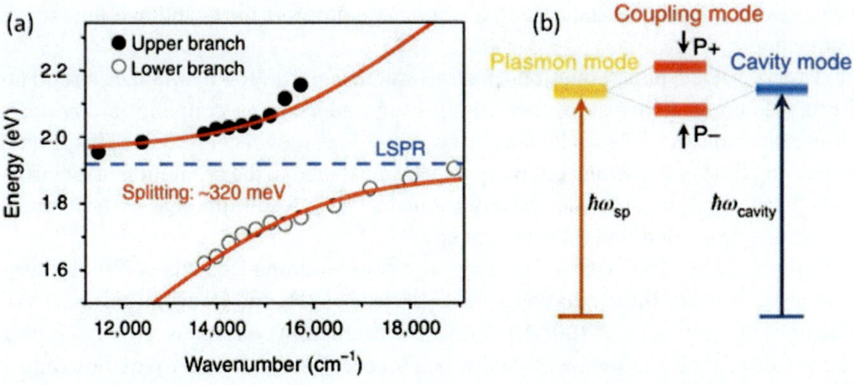

Fig. 3.18 **a** Dispersion curve of the ATA structures. **b** Energy diagram under the strong coupling between LSPR of Au NP and the cavity resonance. Reprinted from [49] Copyright © 2018, Springer Nature

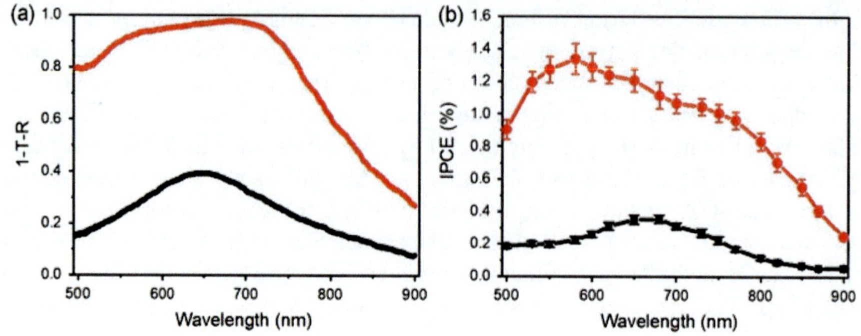

Fig. 3.19 **a** Absorption spectra and **b** IPCE action spectra of ATA and Au NP/TiO_2 with Au NP inlaid depth of 7 nm. Adapted from [49] Copyright © 2018, Springer Nature

PEC measurements were performed using a three-electrode system to determine the light-harvesting performance of the strongly coupled photoanodes. Figure 3.19 shows the absorption and IPCE spectra of ATA and Au NP/TiO_2 with a Au NP inlaid depth of 7 nm. Note that the absorption was also measured under the same environment as the IPCE measurement in a three-electrode system. The IPCE bands show a trend similar to that of the absorption spectra. The dual-band IPCE spectrum of ATA indicates that the lower branch, as well as the upper branch of the strong coupling, can induce charge separation at the Au NP/TiO_2 interface. Note that the photocurrent generation was enhanced in both wavelength regions of the upper and lower branches.

The internal quantum efficiency (IQE), which is obtained by dividing the IPCE value by the absorption, was further investigated to understand the strong coupling effect on the chemical reactions, as shown in Fig. 3.20. As determined by comparing the IQE at the peak wavelength, ATA showed an IQE enhancement of 1.5 times greater than that of Au NP/TiO_2 without strong coupling at both the upper and lower branches. This indicates that strong coupling can promote the quantum efficiency of photochemical reactions.

Time-resolved photoemission electron microscopy (TR-PEEM) measurements further demonstrated that the dephasing time under strong coupling is obviously shorter than that of LSPR under normal conditions, as shown in Fig. 3.21. This shorter dephasing time under strong coupling may be attributed to the promotion of electron-hole pair formation and faster electron transfer. As a result, the IQE of ATA under strong coupling conditions was enhanced.

Two-electrode PEC water splitting was also examined on the ATA structure. Figure 3.22a shows the irradiation time dependence of H_2 and O_2 evolution under visible light irradiation ($\lambda > 400$ nm). Under the irradiation conditions, H_2 and O_2 evolution from the ATA system was stable and proceeded linearly with irradiation time at rates up to 90 nmol/h/W and 43 nmol/h/W, respectively. The H_2 evolution was much higher than that of the uncoupled Au NP/TiO_2 photoelectrode (measured to be 14.3 nmol/h/W) at the same Au NP inlaid depth of 7 nm and under the same experimental conditions. The quantity of H_2 evolved from the Pt wire was nearly twice the quan-

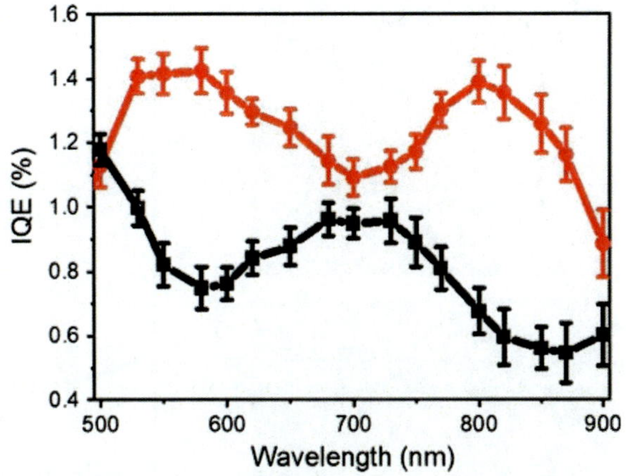

Fig. 3.20 IQE spectra of ATA and Au NP/TiO_2. Adapted from [49] Copyright © 2018, Springer Nature

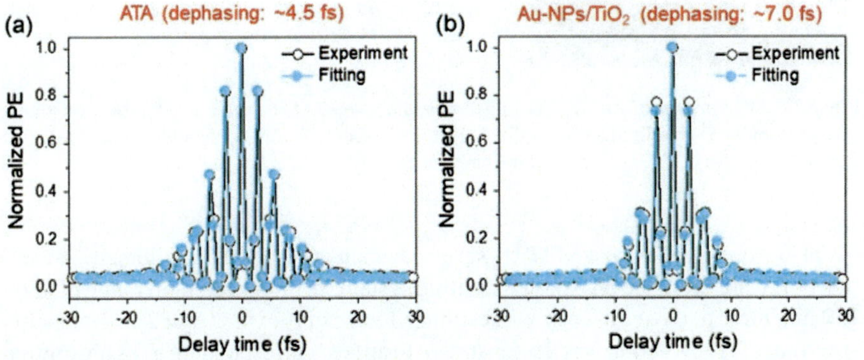

Fig. 3.21 The dephasing time of Au NP under strong coupling conditions **a** and normal conditions **b**. Adapted from [49] Copyright © 2018, Springer Nature

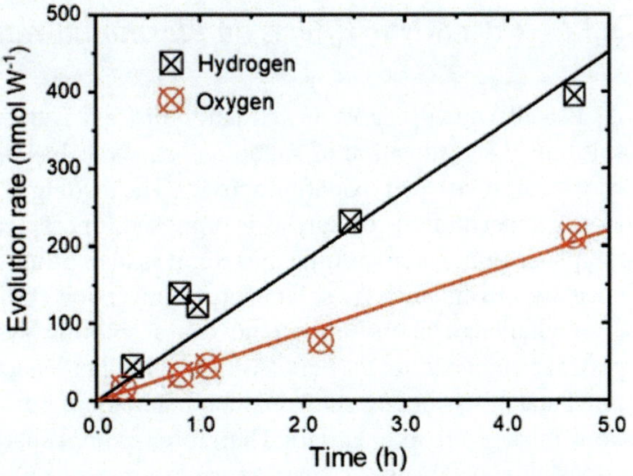

Fig. 3.22 Hydrogen and oxygen evolution of ATA photoelectrode measured in a two-electrode system. Adapted from [49] Copyright © 2018, Springer Nature

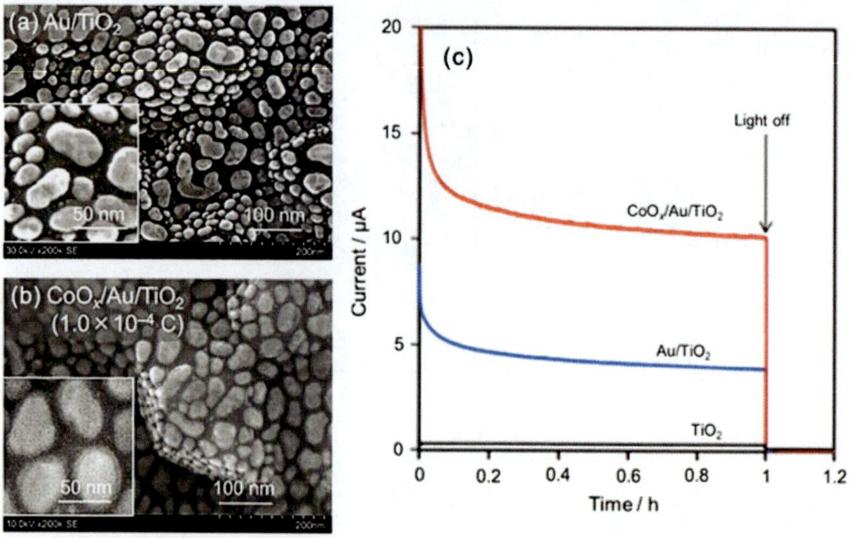

Fig. 3.23 SEM images of Au NP/TiO_2 **a** before and **b** after CoO_x deposition. **c** Photocurrent measurement under visible irradiation on the Au/TiO_2 and CoO_x/Au/TiO_2 photoelectrodes. Adapted from [74] Copyright © 2020, American Chemical Society

tity of O_2 formed on the Au NP/TiO_2 surface, which exhibited a stoichiometric evolution of H_2 and O_2 in the ATA water splitting system. In the wavelength range of 450–750 nm, the Faraday efficiency was estimated to be approximately 82%. The action spectrum of H_2 evolution was demonstrated to almost corresponding to its absorption spectrum, solidly confirming the origination of plasmonic water splitting.

3.3.6 Cocatalytic Effects on Plasmonic Water Splitting

As is well known, the water oxidation process is the bottleneck of complete water splitting. The utilization of suitable cocatalysts has been recognized as an efficient way to improve water oxidation activity. The loading of Mn_3O_4, RuO_2, CoO, and IrO_2 as oxygen evolution cocatalysts is proposed for efficient water oxidation. [70–73] In the plasmonic water splitting system, it is also promising to load a water oxidation cocatalyst to enhance the solar energy conversion efficiency.

A challenge in the plasmonic water splitting system is the short lifetime of plasmon-induced hot carriers. Additionally, hot electrons injected into the conduction band of its support semiconductor should migrate externally without back reactions (charge recombination). Therefore, prompt spatial separation of charge carriers using cocatalysts is a feasible way to improve plasmon-induced charge separation. As chemical reactions are generally considered to take place at the active sites at the boundary between metal NPs and semiconductors, precisely loading the cocat-

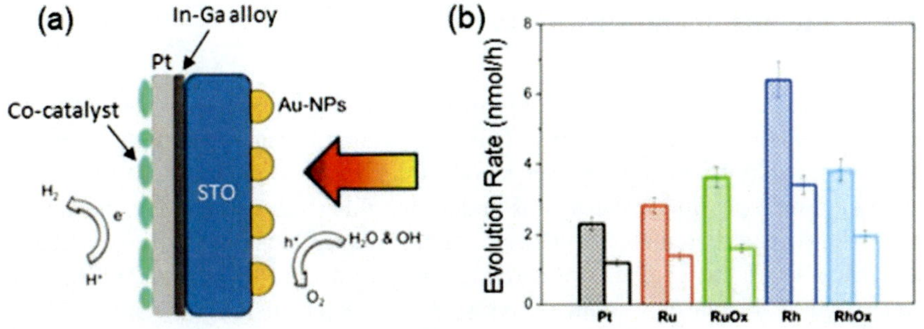

Fig. 3.24 **a** Schematic illustration of water splitting reaction on Au NP/STO loaded with cocatalysts. **b** The rate of H_2 evolution on Au NP/STO loaded with Ru, Rh, and their oxides. Adapted from [77] Copyright © 2015, American Chemical Society

alyst on the active reaction sites is the most effective way to promote charge separation and chemical reactions. On the basis of this concept, Okazaki et al. deposited CoO_x as a water oxidation cocatalyst at the edge of the Au NP/TiO_2 boundary by a photoassisted electrodeposition technique to enhance the plasmonic PEC water oxidation performance [74]. An Au NP-decorated TiO_2 photoanode was fabricated on an FTO substrate with a 30-nm thick TiO_2 layer. The Au NPs were fabricated by the Au thin film annealing method. The CoO_x cocatalyst was loaded on the Au/TiO_2 thin film by a PEC deposition method. CoO_x was mainly deposited at the edge of the Au NP, as shown in Fig. 3.23a, b. The thickness of the CoO_x was estimated to be approximately 6 10 nm, depending on the total amount of reaction charges. From the PEC measurement, the photocurrent generated on the CoO_x/Au/TiO_2 photoanode under visible irradiation ($\lambda > 450$ nm, 180 mW) was enhanced threefold compared to that of Au/TiO_2 without the cocatalyst, as shown in Fig. 3.23c. The plasmonic water oxidation was enhanced by selectively depositing the CoO_x cocatalyst at the active sites of the Au NP/TiO_2 boundary.

To optimize the cocatalytic effect on plasmonic water splitting, several hydrogen evolution cocatalysts were also employed [75, 76]. Based on the self-assisted plasmonic water splitting system discussed above, in the author's group, the hydrogen evolution cocatalytic effect was elucidated by depositing ruthenium or rhodium and their oxide on the platinum board with a thickness of 3–4 nm by sputtering, as the schematic illustration shows in Fig. 3.24a [77]. Figure 3.24b shows the gas evolution rate with various cocatalysts. Platinum and platinum-based nanocomposites have long been considered the most efficient hydrogen evolution catalysts because of their low Gibbs free energy of hydrogen adsorption. In this study, the water splitting efficiency was further enhanced upon loading with Ru, Rh, and their oxides. This further improvement in PEC performance can be explained by the increase in the surface area of the cocatalyst after loading with cocatalysts. Furthermore, a synergetic effect between Pt and the loaded cocatalysts might also contribute to the improvement of the hydrogen evolution reaction.

3.4 Conclusion and Prospects

As numerous experiments have demonstrated, employing plasmonic metal NPs can efficiently address the issues of insufficient utilization of solar energy on semiconductors, especially the wide-bandgap metal oxides that are commonly used for water splitting. Some possible mechanisms underlying this enhancement have been proposed; however, they remain inconsistent. Among tremendous plasmonic photochemical reactions, in this chapter, we introduce the plasmonic PEC water splitting process based on the plasmon-induced charge transfer mechanism. These studies illustrate that plasmonic metallic NP is a promising element that can convert abundant visible to NIR solar energy into chemical energy. Some possible approaches to improve the solar energy conversion efficiency are reviewed. Constructing a metal NP-loaded semiconductor under strong coupling conditions is an efficient way to promote the water splitting reaction in terms of broadening the light absorption and enhancing the water oxidation quantum efficiency. The three-dimensional architecture shows the potential to increase the active sites for plasmonic chemical reactions. Interfacial structure engineering and cocatalysts decorating provide a great opportunity to efficiently modulate plasmon-excited hot-carrier reaction processes for photochemical applications.

Plasmonic chemistry is still in its infancy of development. Although several interesting photochemical processes, such as water splitting, CO_2 reduction, and ammonia photosynthesis, have been successfully demonstrated on metal NP-loaded semiconductors [78–80], their integral solar energy conversion efficiency remains low. There is much room for progress, as well as many challenges, facing the improvement in solar energy conversion efficiency to boost their competitiveness against conventional fossil fuel and silicon-based solar energy conversion systems.

Currently, the mechanism of plasmonic chemical reactions is still under debate, and further fundamental studies are still required. To understand the underlying physics of plasmonic chemical reactions, it is valuable to map out the pathway of plasmon-excited charges. Ultrafast transient spectroscopy provides us with the possibility of measuring the dynamics of the photoexcited charges. Several experiments have been carried out to determine its dynamics by using femtosecond transient absorption spectroscopy [25, 81]. The time scale of plasmon-excited electron transfer from metal NPs to semiconductors has been shown to be less than 50 fs [82]. To map out the complete pathway of plasmon-induced charge separation, not only the dynamics of plasmon-excited electrons, but also plasmon-excited holes need to be clarified. Since the photoexcited hole trapped in the semiconductor generally exhibits absorption maximum in the visible region, which might partially overlap with the plasmon resonance, it is still challenging to measure the dynamics of plasmon-induced holes by transient spectroscopy. Elaborate designs of the transient measurement in future studies might broaden our understanding.

The energy distribution of the plasmon-induced energetic charge carriers, which determines the chemical reactivity of the energetic charge carriers, is another essential parameter for a better understanding of the underlying mechanism. Because the lifetime of the energetic carriers is truly short, generally on the order of 100 fs, it is a major challenge to determine its energy distribution. Recently, photoemission electron microscopy (PEEM) combined with time-resolved pump-probe technology has been used to solve this issue. The high spatial resolution of PEEM gives us precise spatial information of the energetic carriers, and time-resolved technology can help us unfold the energy distribution as well as the dynamics of this energy distribution.

In the future, in addition to noble metal NPs, suitable nonnoble metal NPs could be crafted for plasmonic chemical reactions. Nonnoble metals, such as Al and Cu, that show distinct plasmon resonance in the visible and NIR regions will dramatically reduce the cost of plasmonic energy conversion systems and increase their competitiveness against traditional methods. A stumbling block hindering the use of these low-cost nonnoble metal NPs for efficient plasmonic energy conversion is their stability. Natural oxidation is likely to occur on these nonnoble metal NPs. Looking for a new strategy to use nonnoble metal NPs is one of the promising pathways for the future design of plasmonic energy conversion systems.

To date, most constructions of the plasmonic solar-to-fuel energy conversion system have been photoanodes, where the oxidation reaction takes place. As is widely believed, energetic electrons under plasmon excitation have a broad energy distribution. Efficient utilization of these energetic electrons is also a research interest. In a plasmonic photoanode, the plasmon-excited energetic electrons will finally relax to the conduction band of the semiconductor, losing their energy as heat. Alternatively, when developing a plasmonic photocathode, energetic electrons can transfer directly to the coupling LUMO of molecules without energy loss. Several plasmonic photocathodes have been developed for CO_2 reduction and water splitting [83, 84], although their efficiency is still low. We believe that the future development of the plasmonic photocathode will widely expand the practical applications of plasmonic chemistry.

Acknowledgements The authors would like to thank all contributors of the studies, especially Profs. K. Ueno, K. Murakoshi (Faculty of Science, Hokkaido Univ.), K. Maeda (Department of Chemistry, Tokyo Institute of Technology), Y. Nishijima (Yokohama National Univ.), Q. Sun (Peking University Yangtze Delta Institute of Optoelectronics) and T. Oshikiri (Research Institute for Electronic Science, Hokkaido Univ.) for their valuable collaborations and inspiring discussions. Some of the work described in this article were supported by Nanotechnology Platform (Hokkaido University), and the Dynamic Alliance for Open Innovation Bridging Human, Environment and Materials (Five-Star Alliance) of MEXT.

References

1. S.A. Klein, W.A. Beckman, J.A. Duffie, A design procedure for solar heating systems. Solar Energy **18**, 113–127 (1976)
2. A. Luque, S. Hegedus, *Handbook of Photovoltaic Science and Engineering* (Wiley, Chichester, 2003)
3. Solar splitting of water to hydrogen and oxygen, Allen J. Bard and Marye Anne Fox. Artificial photosynthesis. Acc. Chem. Res. **28**, 141–145 (1995)

4. Y. Tachibana, L. Vayssieres, J.R. Durrant, Artificial photosynthesis for solar water-splitting. Nat. Photon. **6**, 511–518 (2012)
5. A. Fujishima, K. Honda, Electrochemical photolysis of water at a semiconductor electrode. Nature **238**, 37–38 (1972)
6. T. Hisatomi, J. Kubota, K. Domen, Recent advances in semiconductors for photocatalytic and photoelectrochemical water splitting. Chem. Soc. Rev. **43**, 7520–7535 (2014)
7. Meng Ni, Michael K. H. Leung, Dennis Y. C. Leung, and K. Sumathy. A review and recent developments in photocatalytic water-splitting using TiO_2 for hydrogen production. Renew. Sust. Energ. Rev. **11**, 401–425 (2007)
8. Shahed U. M. Khan, Mofareh Al-Shahry, and William B. Ingler. Efficient photochemical water splitting by a chemically modified n-TiO_2. Science **297**, 2243–2245 (2002)
9. O. Teruhisa, M. Takahiro, M. Michio, Photocatalytic activity of S-doped TiO_2 photocatalyst under visible light. Chem. Lett. **32**, 364–365 (2003)
10. Jong Hyeok Park, Sungwook Kim, and Allen J. Bard. Novel carbon-doped TiO_2 nanotube arrays with high aspect ratios for efficient solar water splitting. Nano Lett. 6, **24–28** (2006)
11. T. Umebayashi, T. Yamaki, H. Itoh, K. Asai, Analysis of electronic structures of 3d transition metal-doped TiO_2 based on band calculations. J. Phys. Chem. Solids **63**, 1909–1920 (2002)
12. A. Naldoni, M. Allieta, S. Santangelo, M. Marelli, F. Fabbri, S. Cappelli, C.L. Bianchi, R. Psaro, V. Dal Santo, Effect of nature and location of defects on bandgap narrowing in black TiO_2 nanoparticles. J. Am. Chem. Soc. **134**, 7600–7603 (2012)
13. M. Grätzel, Dye-sensitized solar cells. J. Photochem. Photobiol. C **4**, 145–153 (2003)
14. S. Linic, P. Christopher, D.B. Ingram, Plasmonic-metal nanostructures for efficient conversion of solar to chemical energy. Nat. Mater. **10**, 911–921 (2011)
15. H.A. Atwater, A. Polman, Plasmonics for improved photovoltaic devices. Nat. Mater. **9**, 205–213 (2010)
16. Kyeong-Seok. Lee, Mostafa A. El-Sayed, Gold and silver nanoparticles in sensing and imaging sensitivity of plasmon response to size, shape, and metal composition. J. Phys. Chem. B **110**, 19220–19225 (2006)
17. J. Bosbach, C. Hendrich, F. Stietz, T. Vartanyan, F. Träger, Ultrafast dephasing of surface plasmon excitation in silver nanoparticles: Influence of particle size, shape, and chemical surrounding. Phys. Rev. Lett. **89**, 257404 (2002)
18. K. Kneipp, Y. Wang, H. Kneipp, L.T. Perelman, I. Itzkan, R.R. Dasari, M.S. Feld, Single molecule detection using surface-enhanced Raman scattering (SERS). Phys. Rev. Lett. **78**, 1667–1670 (1997)
19. A. Otto, I. Mrozek, H. Grabhorn, W. Akemann, Surface-enhanced raman scattering. J. Phys. Condens. Matter **4**, 1143–1212 (1992)
20. K. Saha, S.S. Agasti, C. Kim, X. Li, V.M. Rotello, Gold nanoparticles in chemical and biological sensing. Chem. Rev. **112**, 2739–2779 (2012)
21. H.J. Simon, D.E. Mitchell, J.G. Watson, Optical second-harmonic generation with surface plasmons in silver films. Phys. Rev. Lett. **33**, 1531–1534 (1974)
22. Martti Kauranen, Anatoly V. Zayats, Nonlinear plasmonics. Nat. Photon. **6**, 737–748 (2012)
23. K. Ueno, T. Oshikiri, Q. Sun, X. Shi, H. Misawa, Solid-state plasmonic solar cells. Chem. Rev. **118**, 2955–2993 (2018)
24. Y. Tian, T. Tatsuma, Mechanisms and applications of plasmon-induced charge separation at TiO_2 films loaded with gold nanoparticles. J. Am. Chem. Soc. **127**, 7632–7637 (2005)
25. A. Furube, D. Luchao, K. Hara, R. Katoh, M. Tachiya, Ultrafast plasmon-induced electron transfer from gold nanodots into TiO_2 nanoparticles. J. Am. Chem. Soc. **129**, 14852–14853 (2007)
26. J. Li, S.K. Cushing, F. Meng, T.R. Senty, A.D. Bristow, W. Nianqiang, Plasmon-induced resonance energy transfer for solar energy conversion. Nat. Photon. **9**, 601–607 (2015)
27. M.L. Brongersma, N.J. Halas, P. Nordlander, Plasmon-induced hot carrier science and technology. Nat. Nanotechnol. **10**, 25–34 (2015)
28. S.C. Warren, E. Thimsen, Plasmonic solar water splitting. Energy Environ. Sci. **5**, 5133–5146 (2012)

29. Yuchao Zhang, Shuai He, Wenxiao Guo, Yue Hu, Jiawei Huang, Justin R. Mulcahy, and Wei David Wei. Surface-plasmon-driven hot electron photochemistry. Chem. Rev. **118**, 2927–2954 (2018)
30. Z. Bian, T. Tachikawa, P. Zhang, M. Fujitsuka, T. Majima, Au/TiO_2 superstructure-based plasmonic photocatalysts exhibiting efficient charge separation and unprecedented activity. J. Am. Chem. Soc. **136**, 458–465 (2014)
31. Y. Nishijima, K. Ueno, Y. Yokota, K. Murakoshi, H. Misawa, Plasmon-assisted photocurrent generation from visible to near-infrared wavelength using a Au-nanorods/TiO_2 electrode. J. Phys. Chem. Lett. **1**, 2031–2036 (2010)
32. S. Mubeen, J. Lee, N. Singh, S. Kramer, G.D. Stucky, M. Moskovits, An autonomous photosynthetic device in which all charge carriers derive from surface plasmons. Nat. Nanotechnol. **8**, 247–251 (2013)
33. Z. Zhang, L. Zhang, M.N. Hedhili, H. Zhang, P. Wang, Plasmonic gold nanocrystals coupled with photonic crystal seamlessly on TiO_2 nanotube photoelectrodes for efficient visible light photoelectrochemical water splitting. Nano Lett. **13**, 14–20 (2013)
34. N. Szydlo, R. Poirier, I-V and C-V characteristics of Au/TiO_2 Schottky diodes. J. Appl. Phys. **51**, 3310–3312 (1980)
35. C. Voisin, N. Del Fatti, D. Christofilos, F. Vallée, Ultrafast electron dynamics and optical nonlinearities in metal nanoparticles. J. Phys. Chem. B **105**, 2264–2280 (2001)
36. G.V. Hartland, Optical studies of dynamics in noble metal nanostructures. Chem. Rev. **111**, 3858–3887 (2011)
37. G. Zhao, H. Kozuka, T. Yoko, Sol-gel preparation and photoelectrochemical properties of TiO_2 films containing Au and Ag metal particles. Thin Solid Films **277**, 147–154 (1996)
38. M.W. Knight, H. Sobhani, P. Nordlander, N.J. Halas, Photodetection with active optical antennas. Science **332**, 702–704 (2011)
39. H. Inouye, K. Tanaka, I. Tanahashi, K. Hirao, Ultrafast dynamics of nonequilibrium electrons in a gold nanoparticle system. Phys. Rev. B **57**, 11334–11340 (1998)
40. K. Wu, J. Chen, J.R. McBride, T. Lian, Efficient hot-electron transfer by a plasmon-induced interfacial charge-transfer transition. Science **349**, 632–635 (2015)
41. P. Neutens, P. Van Dorpe, I. De Vlaminck, L. Lagae, G. Borghs, Electrical detection of confined gap plasmons in metal-insulator-metal waveguides. Nat. Photon. **3**, 283–286 (2009)
42. I. Thomann, B.A. Pinaud, Z.B. Chen, B.M. Clemens, T.F. Jaramillo, M.L. Brongersma, Plasmon enhanced solar-to-fuel energy conversion. Nano Lett. **11**, 3440–3446 (2011)
43. Run Shi, Yinhu Cao, Yanjun Bao, Yufei Zhao, Geoffrey I. N. Waterhouse, Zheyu Fang, Li-Zhu Wu, Chen-Ho Tung, Yadong Yin, and Tierui Zhang. Self-assembled Au/CdSe nanocrystal clusters for plasmon-mediated photocatalytic hydrogen evolution. Adv. Mater. **29**, 1700803 (2017)
44. Zhi Wei Seh, Shuhua Liu, Michelle Low, Shuang-Yuan Zhang, Zhaolin Liu, Adnen Mlayah, and Ming-Yong Han. Janus Au-TiO_2 Photocatalysts with Strong Localization of Plasmonic Near-Fields for Efficient Visible-Light Hydrogen Generation. Adv. Mater. **24**, 2310–2314 (2012)
45. Jiun-Jen Chen, Jeffrey C. S. Wu, Pin Chieh Wu, and Din Ping Tsai. Plasmonic photocatalyst for H_2 evolution in photocatalytic water splitting. J. Phys. Chem. C **115**, 210–216 (2010)
46. Y. Nishijima, K. Ueno, Y. Kotake, K. Murakoshi, H. Inoue, H. Misawa, Near-infrared plasmon-assisted water oxidation. J. Phys. Chem. Lett. **3**, 1248–1252 (2012)
47. Z. Liu, W. Hou, P. Pavaskar, M. Aykol, S.B. Cronin, Plasmon resonant enhancement of photocatalytic water splitting under visible illumination. Nano Lett. **11**, 1111–1116 (2011)
48. J. LeeJ, S. Mubeen, X. Ji, G.D. Stucky, M. Moskovits, Plasmonic photoanodes for solar water splitting with visible light. Nano Lett. **12**, 5014–5019 (2012)
49. X. Shi, K. Ueno, T. Oshikiri, Q. Sun, K. Sasaki, H. Misawa, Enhanced water splitting under modal strong coupling conditions. Nat. Nanotechnol. **13**, 953–958 (2018)
50. S.Y. Chou, P.R. Krauss, P.J. Renstrom, Imprint of sub]25 nm vias and trenches in polymers. Appl. Phys. Lett. **67**, 3114–3116 (1995)

51. K. Ueno, S. Juodkazis, V. Mizeikis, K. Sasaki, H. Misawa, Spectrally-resolved atomic-scale length variations of gold nanorods. J. Am. Chem. Soc. **128**, 14226–14227 (2006)
52. H. Lang, S. Maldonado, K.J. Stevenson, B.D. Chandler, Synthesis and characterization of dendrimer templated supported bimetallic Pt-Au nanoparticles. J. Am. Chem. Soc. **126**, 12949–12956 (2004)
53. Q. Sun, Yu. Han, K. Ueno, A. Kubo, Y. Matsuo, H. Misawa, Dissecting the few-femtosecond dephasing time of dipole and quadrupole modes in gold nanoparticles using polarized photoemission electron microscopy. ACS Nano **10**, 3835–3842 (2016)
54. Yu. Han, Q. Sun, K. Ueno, T. Oshikiri, A. Kubo, Y. Matsuo, H. Misawa, Exploring coupled plasmonic nanostructures in the near field by photoemission electron microscopy. ACS Nano **10**, 10373–10381 (2016)
55. Michael K. L. Man, Athanasios Margiolakis, Skylar Deckoff-Jones, Takaaki Harada, E. Laine Wong, M. Bala Murali Krishna, Julien Madéo, Andrew Winchester, Sidong Lei, Robert Vajtai, Pulickel M. Ajayan, and Keshav M. Dani. Imaging the motion of electrons across semiconductor heterojunctions. Nat. Nanotechnol. **12**, 36–40 (2017)
56. Jinse Park, P. R. Deshmukh, Youngku Sohn, and Weon Gyu Shin. ZnO-TiO_2 core-shell nanowires decorated with Au nanoparticles for plasmon-enhanced photoelectrochemical water splitting. J. Alloy Compd. **787**, 1310–1319 (2019)
57. H. LiH, Z. Li, Yu. Yanhao, Y. Ma, W. Yang, F. Wang, X. Yin, X. Wang, Surface-plasmon-resonance-enhanced photoelectrochemical water splitting from Au-nanoparticle-decorated 3D TiO_2 nanorod architectures. J. Phys. Chem. C **121**, 12071–12079 (2017)
58. X. Shi, K. Ueno, N. Takabayashi, H. Misawa, Plasmon-enhanced photocurrent generation and water oxidation with a gold nanoisland-loaded titanium dioxide photoelectrode. J. Phys. Chem. C **117**, 2494–2499 (2012)
59. V. Fthenakis, Sustainability of photovoltaics: The case for thin-film solar cells. Renew. Sust. Energ. Rev. **13**, 2746–2750 (2009)
60. K. Sunada, Y. Kikuchi, K. Hashimoto, A. Fujishima, Bactericidal and detoxification effects of TiO_2 thin film photocatalysts. Environ. Sci. Technol. **32**, 726–728 (1998)
61. Makiko Yamagishi, Shna Kuriki, P. K. Song, and Yuzo Shigesato. Thin film TiO_2 photocatalyst deposited by reactive magnetron sputtering. Thin Solid Films **442**, 227–231 (2003)
62. X. Shi, K. Ueno, T. Oshikiri, H. Misawa, Improvement of plasmon-enhanced photocurrent generation by interference of TiO_2 thin film. J. Phys. Chem. C **117**, 24733–24739 (2013)
63. X. Shi, X. Li, T. Toda, T. Oshikiri, K. Ueno, K. Suzuki, K. Murakoshi, H. Misawa, Interfacial structure-modulated plasmon-induced water oxidation on strontium titanate. ACS Appl. Energy Mater. **3**, 5675–5683 (2020)
64. Y. Zhong, K. Ueno, X. Yuko Mori, T.O. Shi, K. Murakoshi, H. Inoue, H. Misawa, Plasmon-assisted water splitting using two sides of the same $SrTiO_3$ single-crystal substrate: conversion of visible light to chemical energy. Angew. Chem. Int. Ed. **53**, 1–6 (2014)
65. Y.-C. Yen, J.-A. Chen, O. Sheng, Y.-S. Chen, K.-J. Lin, Plasmon-enhanced photocurrent using gold nanoparticles on a three-dimensional TiO_2 nanowire-web electrode. Sci. Rep. **7**, 42524 (2017)
66. R. Takakura, T. Oshikiri, X. Kosei Ueno, T.K. Shi, H. Masuda, H. Misawa, Water splitting using a three-dimensional plasmonic photoanode with titanium dioxide nano-tunnels. Green Chemistry **19**, 2398–2405 (2017)
67. E. Orgiu, J. George, J.A. Hutchison, E. Devaux, J.F. Dayen, B. Doudin, F. Stellacci, C. Genet, J. Schachenmayer, C. Genes, G. Pupillo, P. Samorì, T.W. Ebbesen, Conductivity in organic semiconductors hybridized with the vacuum field. Nat. Mater. **14**, 1123–1129 (2015)
68. J.A. Hutchison, T. Schwartz, C. Genet, E. Devaux, T.W. Ebbesen, Modifying chemical landscapes by coupling to vacuum fields. Angew. Chem. Int. Ed. **51**, 1592–1596 (2012)
69. T. Schwartz, J.A. Hutchison, C. Genet, T.W. Ebbesen, Reversible switching of ultrastrong light-molecule coupling. Phys. Rev. Lett. **106**, 196405 (2011)
70. A. Ramírez, P. Hillebrand, D. Stellmach, M.M. May, P. Bogdanoff, S. Fiechter, Evaluation of MnO_x, Mn_2O_3, and Mn_3O_4 electrodeposited films for the oxygen evolution reaction of water. J. Phys. Chem. C **118**, 14073–14081 (2014)

71. S. Licht, B. Wang, S. Mukerji, T. Soga, M. Umeno, H. Tributsch, Efficient solar water splitting, exemplified by RuO_2-catalyzed AlGaAs/Si photoelectrolysis. J. Phys. Chem. B **104**, 8920–8924 (2000)
72. L. Liao, Q. Zhang, S. Zhihua, Z. Zhao, Y. Wang, Y. Li, L. Xiaoxiang, D. Wei, G. Feng, Yu. Qingkai, X. Cai, J. Zhao, Z. Ren, H. Fang, F. Robles-Hernandez, S. Baldelli, J. Bao, Efficient solar water-splitting using a nanocrystalline CoO photocatalyst. Nat. Nanotechnol. **9**, 69–73 (2014)
73. R. Asai, H. Nemoto, Q. Jia, K. Saito, A. Iwase, A. Kudo, A visible light responsive rhodium and antimony-codoped $SrTiO_3$ powdered photocatalyst loaded with an IrO_2 cocatalyst for solar water splitting. Chem. Commun. **50**, 2543–2546 (2014)
74. M. Okazaki, Y. Suganami, N. Hirayama, H. Nakata, T. Oshikiri, T. Yokoi, H. Misawa, K. Maeda, Site-selective deposition of a cobalt cocatalyst onto a plasmonic Au/TiO_2 photoanode for improved water oxidation. ACS Appl. Energy Mater. **3**, 5142–5146 (2020)
75. G.P. Singh, K.M. Shrestha, A. Nepal, K.J. Klabunde, C.M. Sorensen, Graphene supported plasmonic photocatalyst for hydrogen evolution in photocatalytic water splitting. Nanotechnology **25**, 265701 (2014)
76. A. Tanaka, K. Teramura, S. Hosokawa, H. Kominami, T. Tanaka, Visible light-induced water splitting in an aqueous suspension of a plasmonic Au/TiO_2 photocatalyst with metal co-catalysts. Chem. Sci. **8**, 2574–2580 (2017)
77. Y. Zhong, K. Ueno, Y. Mori, T. Oshikiri, H. Misawa, Cocatalyst effects on hydrogen evolution in a plasmon-induced water-splitting system. J. Phys. Chem. C **119**, 8889–8897 (2015)
78. Wenbo Hou, Wei Hsuan Hung, Prathamesh Pavaskar, Alain Goeppert, Mehmet Aykol, and Stephen B. Cronin. Photocatalytic conversion of CO_2 to hydrocarbon fuels via plasmon-enhanced absorption and metallic interband transitions. ACS Catal. **1**, 929–936 (2011)
79. T. Oshikiri, K. Ueno, H. Misawa, Plasmon-induced ammonia synthesis through nitrogen photofixation with visible light irradiation. Angew. Chem. Int. Ed. **53**, 9802–9805 (2014)
80. T. Oshikiri, K. Ueno, H. Misawa, Selective dinitrogen conversion to ammonia using water and visible light through plasmon-induced charge separation. Angew. Chem. Int. Ed. **55**, 3942–3946 (2016)
81. Hayk Harutyunyan, Alex B. F. Martinson, Daniel Rosenmann, Larousse Khosravi Khorashad, Lucas V. Besteiro, Alexander O. Govorov, and Gary P. Wiederrecht. Anomalous ultrafast dynamics of hot plasmonic electrons in nanostructures with hot spots. Nat. Nanotechnol. **10**, 770–774 (2015)
82. elucidation of the electron eobility before deep trapping, Yoshiaki Tamaki, Kohjiro Hara, Ryuzi Katoh, M. Tachiya, and Akihiro Furube. Femtosecond visible-to-IR spectroscopy of TiO_2 nanocrystalline films. J. Phys. Chem. C **113**, 11741–11746 (2009)
83. J.S. DuChene, G. Tagliabue, A.J. Welch, W.-H. Cheng, H.A. Atwater, Hot hole collection and photoelectrochemical CO_2 reduction with plasmonic Au/p-GaN photocathodes. Nano Lett. **18**, 2545–2550 (2018)
84. D. ChenD, Z. Liu, Z. Guo, W. Yan, Y. Xin, Enhancing light harvesting and charge separation of Cu_2O photocathodes with spatially separated noble-metal cocatalysts towards highly efficient water splitting. J. Mater. Chem. A **6**, 20393–20401 (2018)

Chapter 4
Ab Initio Computational Approach for Nanophotonics Based on Time-Dependent Density Functional Theory

Kazuhiro Yabana, Takashi Takeuchi, Mitsuharu Uemoto, Atsushi Yamada, and Shunsuke Yamada

Abstract We present theoretical and computational approaches to describe ultrafast and nonlinear optical responses in nano-materials based on ab initio time-dependent density functional theory. The method is applicable to a wide variety of phenomena in nanophotonics, including nonlinear optical responses of thin films and metasurfaces, and coherent phonon generations.

4.1 Introduction

Nowadays ab initio computational approaches represented by density functional theory are widely used and recognized as indispensable tools in many fields of physics, chemistry, and material sciences. In this article, we present our attempt to develop a theoretical and computational approach to explore ultrafast and nonlinear optical responses of nano-sized materials based on time-dependent density functional theory (TDDFT) [1, 2].

K. Yabana (✉) · T. Takeuchi · A. Yamada · S. Yamada
Center for Computational Sciences, University of Tsukuba,Tsukuba, Japan
e-mail: yabana@nucl.ph.tsukuba.ac.jp

T. Takeuchi
e-mail: take@ccs.tsukuba.ac.jp

A. Yamada
e-mail: ayamada@ccs.tsukuba.ac.jp

S. Yamada
e-mail: syamada@ccs.tsukuba.ac.jp

M. Uemoto
Department of Electrical and Electronic Engineering, Kobe University, Kobe, Japan
e-mail: uemoto@eedept.kobe-u.ac.jp

T. Yatsui (ed.), *Progress in Nanophotonics 6*, Nano-Optics and Nanophotonics,
https://doi.org/10.1007/978-3-030-71516-8_4

In optical science, two kinds of computational approaches have been developed so far. One is ab initio quantum mechanical approaches to evaluate optical parameters such as dielectric function, calculating electronic responses to an applied field using perturbation theories. The other is electromagnetism analyses solving Maxwell equations for a given geometry of materials where optical constants of the media are treated as given parameters. These two approaches assume two aspects in the light-matter interaction: weakness of the light field and difference in the length scale that guarantees the use of the local and linear constitutive relation of the macroscopic electromagnetism. However, the usefulness and capability of these traditional computational approaches are limited in the current frontiers of nanophotonics.

In nano-sized materials, the electronic structure at the surface becomes important and is different from the bulk medium. In particular, it is apparent that we cannot apply ordinary macroscopic description for two-dimensional (2D) atomic layer materials. It indicates that it is important to treat the nonlocality in the optical response of nano-sized materials [3]. Using intense and ultrashort laser pulses, extremely nonlinear optical responses have been extensively investigated for nano-materials as well as atoms, molecules, and solids [4]. In these circumstances, ordinary descriptions using local and linear constitutive relations are no more useful. We consider that a unified computational approach that combines quantum mechanical description of electrons and ions in microscopic scale and electromagnetism analysis of light fields in macroscopic scale is required.

The TDDFT has been known as a useful method to investigate optical properties of molecules and solids with a reasonable accuracy for a moderate computational cost [1, 2]. Solving the time-dependent Kohn-Sham (TDKS) equation in time domain [5, 6], it is possible to describe ultrafast and nonlinear electronic dynamics in matters induced by a pulsed electric field without any perturbative approximations. For molecules, it has been utilized to calculate photoabsorption [7, 8], nonlinear susceptibilities [9], extremely nonlinear phenomena such as high harmonic generation [10] and Coulomb explosions [11]. For bulk solids, it has been utilized to calculate dielectric functions [6], nonlinear susceptibilities [12], coherent phonon generation [13], optical breakdown of dielectrics [14], high harmonic generations [15], and optically induced current in dielectrics [16]. Using ordinary density functionals, however, electron-electron collision effects cannot be included sufficiently, although carrier relaxation dynamics is known to be important in the time scale as short as a few tens of femtosecond. This limits the validity of the TDDFT approach to a short duration before the relaxation becomes significant.

To describe propagations and scatterings of pulsed light by nano-materials, it is necessary to couple the microscopic dynamics of electrons and ions described by TDDFT with the dynamics of light electromagnetic fields. We note that there are two length scales in the optical response: a typical wavelength of an optical light pulse that is a few hundred nanometer, and the motion of electrons and ions induced by the optical field that is at most an Angström. Depending on which spatial scale is concerned, we have been developing a few schemes of coupled dynamics.

When we describe optical responses of nano-materials whose sizes are comparable to the wavelength of light, a three-dimensional (3D) macroscopic approach will be appropriate. In the method, two different spatial grids are utilized to describe macroscopic light electromagnetic fields and microscopic electron dynamics [17]. We often call it multiscale Maxwell-TDDFT method. This is useful when we describe extremely nonlinear optical responses of nano-materials and bulk surfaces. This multiscale method has been successfully applied to analyze ultrafast and nonlinear light propagations through thin films, for example, mimicking pump-probe experiments of the attosecond transient absorption spectroscopy [18], a spatial distribution of the energy deposition that is the basic information to analyze laser processing [19, 20]. For extremely thin materials, including atomic layered materials, the 2D macroscopic approach has also been developed. Here the electromagnetic fields in the thin film are treated as a spatially uniform field ignoring the propagation of the light. Finally, a microscopic approach in which a common spatial grid is used to solve Maxwell and TDKS equations [21] has been developed. We often call it single-scale Maxwell-TDDFT method. This approach should be applicable for any situation although the maximum size is limited by a practical limit of computation.

In this article, we first describe the theoretical and computational methods described above. First we explain theoretical schemes. For microscopic dynamics of electrons and ions, based on TDDFT for isolated and periodic systems. Coupled theories of the electromagnetism analysis and TDDFT, 3D macroscopic, 2D macroscopic, and single-scale methods will then be explained. We next show some recent applications of the methods. After describing basic features of the electron dynamics calculations for a given applied electric field, we present simultaneous calculations of light propagation and microscopic electron-ion dynamics. For a thin film of Si, we present a comparison of three approaches, microscopic, 2D macroscopic, and 3D macroscopic methods [21]. We also present a combined calculation of phonon dynamics of coherent phonons in a diamond that involves Raman process [22]. We then describe optical responses of metasurfaces. Using the 2D macroscopic method, optical responses of plasmonic metasurfaces are calculated using TDDFT with jellium model [23]. Using the 3D macroscopic method, nonlinear optical responses of metasurfaces of semiconductors are presented [24]. Finally, a summary is presented.

4.2 Formalism

4.2.1 Optical Response in Dipole Approximation

In this section, we explain a description of optical responses of matters using TDDFT in the dipole approximation. Since a typical wavelength of light is much longer than the typical spatial scale of electronic and ionic motions in matters, it is usually sufficient to consider the response to a spatially uniform, time-dependent electric field $\mathbf{E}(t)$.

First, we consider an isolated system such as molecules and nanoparticles composed of N_I ions and N_e valence electrons. When the electric field $\mathbf{E}(t)$ is applied to the system, electron motion is described by the TDKS equation for electron orbitals $\psi_i(\mathbf{r}, t)$

$$i\hbar\frac{\partial}{\partial t}\psi_i(\mathbf{r}, t) = \left\{\frac{\mathbf{p}^2}{2m} + V_{\mathrm{ion}}(\mathbf{r}, \mathbf{S}_a) + \int d\mathbf{r}'\frac{e^2}{|\mathbf{r}-\mathbf{r}'|}n(\mathbf{r}', t) + V_{\mathrm{xc}}[n(\mathbf{r}, t)] + e\mathbf{E}(t)\cdot\mathbf{r}\right\}\psi_i(\mathbf{r}, t), \tag{4.1}$$

where V_{ion} is the potential between electrons and ions for which we utilize norm-conserving pseudopotential [26] in practical calculations. V_{xc} is the exchange-correlation potential. The electron density is given by

$$n(\mathbf{r}, t) = \sum_{i=1}^{N_e}|\psi_i(\mathbf{r}, t)|^2. \tag{4.2}$$

For ions whose coordinates are expressed as $\mathbf{S}_\alpha$, $\alpha = 1, \ldots, N_I$, we assume that they are described by a classical Newtonian dynamics with the force determined by the instantaneous distribution of electrons and other ions

$$M_\alpha\frac{d^2\mathbf{S}_\alpha}{dt^2} = eZ_\alpha\mathbf{E}(t) - \frac{\partial}{\partial\mathbf{S}_\alpha}\left[\sum_{\beta\neq\alpha}\frac{Z_\alpha Z_\beta e^2}{|\mathbf{S}_\alpha - \mathbf{S}_\beta|} - \int d\mathbf{r}\frac{Z_\alpha e^2}{|\mathbf{r}-\mathbf{S}_\alpha|}n(\mathbf{r}, t)\right], \tag{4.3}$$

where Z_α is the valence charge number of α-th ion. This is often called the Ehrenfest molecular dynamics calculation. In practical calculations, we again utilize the norm-conserving pseudopotential for electron-ion interactions.

We next consider a description of an infinite system where the ionic configuration specified by $\mathbf{S}_\alpha$ forms a periodic lattice. To describe electron motion in infinite periodic systems, it is essential to use the Bloch theorem. However, the Hamiltonian in (4.1), which we call the length-gauge Hamiltonian, is not periodic due to the potential of the uniform electric field, $e\mathbf{E}(t)\cdot\mathbf{r}$. To recover the periodicity, we utilize a gauge transformation from the length gauge to the velocity gauge

$$\psi_i(\mathbf{r}, t) = \exp\left[\frac{ie}{\hbar c}\mathbf{A}(t)\cdot\mathbf{r}\right]\tilde{\psi}_i(\mathbf{r}, t), \tag{4.4}$$

where the vector potential $\mathbf{A}(t)$ is related to the electric field $\mathbf{E}(t)$ by

$$\mathbf{E}(t) = -\frac{1}{c}\frac{d\mathbf{A}(t)}{dt}. \tag{4.5}$$

The TDKS equation in the velocity gauge is expressed as

$$i\hbar\frac{\partial}{\partial t}\tilde{\psi}_i(\mathbf{r},t) = \left\{\frac{1}{2m}\left(\mathbf{p}+\frac{e}{c}\mathbf{A}(t)\right)^2 + V_{\text{ion}}(\mathbf{r},\mathbf{S}_a) + \int d\mathbf{r}'\frac{e^2}{|\mathbf{r}-\mathbf{r}'|}n(\mathbf{r}',t) + V_{\text{xc}}[n(\mathbf{r},t)]\right\}\tilde{\psi}_i(\mathbf{r},t), \tag{4.6}$$

Now the Hamiltonian has the same periodicity as the lattice, and we can apply the Bloch theorem. The Bloch orbitals are introduced by

$$\tilde{\psi}_i(\mathbf{r},t) = e^{i\mathbf{k}\cdot\mathbf{r}}u_{b\mathbf{k}}(\mathbf{r},t), \tag{4.7}$$

where the orbital index i is now composed of the band index b and the wave vector $\mathbf{k}$. The TDKS equation for the Bloch orbital is given by

$$i\hbar\frac{\partial}{\partial t}u_{b\mathbf{k}}(\mathbf{r},t) = H_{\mathbf{k}+e\mathbf{A}(t)/\hbar c}u_{b\mathbf{k}}(\mathbf{r},t), \tag{4.8}$$

$$H_{\mathbf{k}} = \frac{1}{2m}(\mathbf{p}+\hbar\mathbf{k})^2 + V_{\text{ion}}(\mathbf{r},\mathbf{S}_a) + \int d\mathbf{r}'\frac{e^2}{|\mathbf{r}-\mathbf{r}'|}n(\mathbf{r}',t) + V_{\text{xc}}[n(\mathbf{r},t)]. \tag{4.9}$$

The important output from the time evolution calculation is the electric current density. The current density from the electronic motion, $\mathbf{j}(\mathbf{r},t)$, is given by

$$\mathbf{j}(\mathbf{r},t) = \frac{1}{2m}\sum_i\left\{u_{b\mathbf{k}}^*(\mathbf{r},t)\left(-i\hbar\nabla+\mathbf{k}+\frac{e}{c}\mathbf{A}(t)\right)u_{b\mathbf{k}}(\mathbf{r},t)+c.c.\right\}, \tag{4.10}$$

Taking the spatial average over the unit cell, we also introduce the electric current density averaged over the unit cell volume, $\mathbf{J}(t)$, by

$$\mathbf{J}(t) = \frac{1}{\Omega}\int d\mathbf{r}\mathbf{j}(\mathbf{r},t) + \mathbf{J}_{ps}(t), \tag{4.11}$$

where Ω is the volume of the unit cell and $\mathbf{J}_{ps}$ is a current contributed from the nonlocal part of the pseudopotential. When we treat the ionic motion as well, we need to include the electric current induced by ions.

4.2.2 *Coupled Description of Electromagnetic Fields*

4.2.2.1 3D Macroscopic Description

We next discuss how we can describe the propagation of pulsed light together with the microscopic dynamics of electrons and ions. We first consider the case in which the object that is irradiated by the pulsed light is spatially so extended that the light propagation may be described employing the macroscopic electromagnetism with a course-grained modeling [17]. In such a situation, we utilize 3D Maxwell-TDDFT multiscale method as described below. We assume that the propagation of the macroscopic electromagnetic fields can be described by the following macroscopic Maxwell equation for the vector potential

$$\nabla \times \nabla \times \mathbf{A}(\mathbf{R}, t) + \frac{1}{c^2}\frac{\partial^2}{\partial t^2}\mathbf{A}(\mathbf{R}, t) = \frac{4\pi}{c}\mathbf{J}(\mathbf{R}, t), \quad (4.12)$$

where we introduced the coordinate $\mathbf{R}$ to express macroscopic quantities. We need to specify the prescription to connect the vector potential $\mathbf{A}(\mathbf{R}, t)$ and the current density $\mathbf{J}(\mathbf{R}, t)$ from the microscopic dynamics.

In the macroscopic modeling, we assume that the current density $\mathbf{J}(\mathbf{R}, t)$ at each position $\mathbf{R}$ is determined by the field at the same position, namely by the vector potential $\mathbf{A}(\mathbf{R}, t)$ by the following procedure. At each position $\mathbf{R}$, we consider a microscopic system for which we assume a uniform approximation. Namely, we treat the microscopic dynamics in the dipole approximation. At each macroscopic position specified by $\mathbf{R}$, we prepare Bloch orbitals, $u_{\mathbf{R},b\mathbf{k}}(\mathbf{r}, t)$, that satisfy

$$i\hbar\frac{\partial}{\partial t}u_{\mathbf{R},b\mathbf{k}}(\mathbf{r}, t) = H_{\mathbf{k}+e\mathbf{A}(\mathbf{R},t)/\hbar c}u_{\mathbf{R},b\mathbf{k}}(\mathbf{r}, t), \quad (4.13)$$

where the coordinate $\mathbf{R}$ is treated as a parameter in solving the above TDKS equation. Ionic coordinates are also prepared at each $\mathbf{R}$, $\mathbf{S}_{\mathbf{R},\alpha}$, and a similar Newtonian equation as (4.3) is solved. The microscopic dynamics provides the electric current density using (4.11) utilizing the Bloch orbitals and ionic coordinates of the position $\mathbf{R}$. In this way, we obtain a closed set of equations that describe the propagation of the macroscopic electromagnetic fields and the microscopic dynamics of electrons and ions, simultaneously. We note that, in practical calculations, it is necessary to solve TDKS equations at many grid points of $\mathbf{R}$ simultaneously. We may utilize parallel supercomputers efficiently for the calculation.

4.2.2.2 2D Macroscopic Description

To describe light propagation through a very thin film or an atomic layered material, we develop a 2D macroscopic approach [21]. This approach is applicable to problems in which the electromagnetic fields inside the film can be regarded as position-independent. The condition is satisfied for a film of thickness less than about 10 nm.

To formulate the method, we consider a thin film material that has a periodicity in xy-plane. We make a 2D course-graining approximation for the microscopic electric current density of (4.10),

$$\mathbf{J}_{2D}(t) = \int \frac{dxdy}{S_{xy}} \int dz\mathbf{j}(\mathbf{r}, t), \quad (4.14)$$

where S_{xy} is the area of the 2D unit cell. We assume that the incident pulse irradiates normally on the thin film and propagates along the z-axis, and utilize the coordinate Z to express the propagation of the light. Assuming that the thickness of the thin film is very small, the macroscopic Maxwell equation is approximated as

$$-\frac{\partial^2}{\partial Z^2}A(Z,t)+\frac{1}{c^2}\frac{\partial^2}{\partial t^2}A(Z,t)=\frac{4\pi}{c}\delta(Z)J_{2D}(t). \tag{4.15}$$

We find a solution of this equation for a given incident pulse specified by the time profile, $A^{in}(t)$. The equation for the transmitted pulse $A^{tr}(t)$ is derived as [21]

$$\frac{dA^{tr}}{dt}=\frac{dA^{in}}{dt}+\frac{2\pi}{c}J_{2D}[A^{tr}](t), \tag{4.16}$$

where $J_{2D}[A^{tr}](t)$ is the 2D course-grained current density induced by the macroscopic electric field of $A^{tr}(t)$. For a weak field, this expression gives the transmission coefficient

$$\frac{E^{tr}(\omega)}{E^{in}(\omega)}=\frac{1}{1+\frac{2\pi}{c}\sigma_{2D}(\omega)}, \tag{4.17}$$

where $\sigma_{2D}(\omega)$ is the 2D conductivity. This formula is often used to investigate optical properties of 2D materials. Equation (4.16) is not limited by the intensity of the pulse and can be used to explore extremely nonlinear responses.

4.2.2.3 3D Microscopic Description

Finally, we implemented a simple and straightforward scheme, just to solve the Maxwell equation together with the TDKS equation using the same spatial grids without course graining [21]. We call it 3D Maxwell-TDDFT single-scale method. The microscopic vector potential $\mathbf{A}(\mathbf{r},t)$ that describes transverse electromagnetic fields satisfies

$$\nabla\times\nabla\times\mathbf{A}(\mathbf{r},t)+\frac{1}{c^2}\frac{\partial^2}{\partial t^2}\mathbf{A}(\mathbf{r},t)=\frac{4\pi}{c}\mathbf{j}^T(\mathbf{r},t), \tag{4.18}$$

where $\mathbf{j}^T$ is the transverse component of the electric current density of (4.10). Solving (4.6) and (4.18) simultaneously, we can describe time evolutions of electromagnetic fields and electronic (and ionic) dynamics. Although this 3D microscopic description is applicable, in principle, to any circumstances, application is in practice limited by the computational feasibility.

4.3 Bulk Solids

Before discussing optical responses of nano-materials, we first show calculations of electron dynamics and optical properties of bulk solids taking silicon (Si) as an example [25]. We take a cubic unit cell containing eight Si atoms. The norm-conserving pseudopotential of Troullier-Martin type [26] is utilized for electron-ion potential. For a reliable description of laser-solid interaction, it is important to utilize energy functional that reproduces energetic features of materials quantitatively. We employ the TB-mBJ potential [27] in adiabatic approximation. The TB-mBJ potential is a meta-GGA potential that depends on density, gradient of the density, and kinetic energy density, and is known to reproduce band gap energies of dielectrics systematically. The TB-mBJ potential includes one parameter c. We determine it to accurately reproduce the energy of the indirect band gap of Si, $c = 1.04$ that gives the indirect band gap energy of 1.26 eV. For the applied electric field, we use the following time profile of cosine-squared envelope shape for the vector potential,

$$A(t) = -\frac{A_0}{\omega} \cos^2\left(\frac{\pi t}{T}\right) \cos \omega t, \qquad (-\frac{T}{2} < t < \frac{T}{2}). \tag{4.19}$$

All calculations presented in this article, we utilize an open-source software SALMON (Scalable Ab initio Light-Matter simulator for Optics and Nanoscience) [28] that is developed by us. In SALMON, 3D grid representation is used to express Kohn-Sham orbitals and other quantities. The time evolution of the Kohn-Sham orbitals is calculated using the scheme developed in [5, 6, 17].

For the calculation of Si, the spatial grid size, as well as numbers of k-points and time step size Δt are determined, so that the calculated results converge. The determined parameters of the grid and step sizes are $\Delta x = 0.34$Å , 12^3 of k-points, and $\Delta t = 0.48$ attosecond(as). The duration of the pulse is set to $T = 15$ fs (5.5 fs at the FWHM of the envelope), while the amplitude A_0 and the frequency ω are varied. To indicate the intensity of the pulsed laser light, we use the maximum intensity I determined from the maximum amplitude of the pulsed electric field. We note that the relation connecting the amplitude of the electric field and the intensity of the pulsed laser light in the vacuum is utilized, not that in the medium.

In Fig. 4.1, we show a typical electron dynamics calculation in a crystalline silicon. In the upper panel, the time profile of the pulsed electric field $E(t)$ (black line) and the induced electric current density $J(t)$ is shown for two cases of different intensities, $I = 10^{11}$ W/cm^2 and $I = 10^{13}$ W/cm^2. Note that the current density at $I = 10^{11}$ W/cm^2 is multiplied by a factor of 10. The frequency of the pulsed electric field is set to $\hbar\omega = 1.55$ eV. At the intensity of $I = 10^{11}$ W/cm^2, the electric current density shows a similar time profile to the pulsed electric field, except for a phase difference of $\pi/2$. This indicates a linear dielectric response of silicon for a weak electric field at the frequency below the direct band gap energy. At the intensity of $I = 10^{13}$ W/cm^2, the time profile of the electric current density is strongly distorted

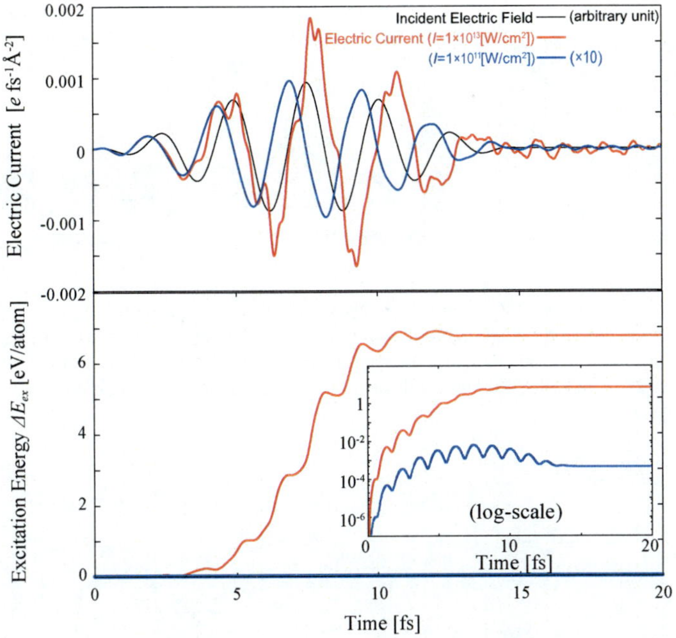

Fig. 4.1 A typical time evolution calculation in Si. In the upper panel, the applied pulsed electric field is shown by a black-solid line. Induced electric current density is shown at two intensities, blue line for $I = 10^{11}$ W/cm^2 (multiplied by a factor 10) and red line for $I = 10^{13}$ W/cm^2. The frequency is set to $\hbar\omega = 1.55eV$. In the lower panel, the electronic excitation energy per atom is shown. Taken from [25] with kind permission of The European Physical Journal (EPJ)

and is very different from the lower intensity case. This indicates a strongly nonlinear response of electrons to the electric field.

In the lower panel, the electronic excitation energy $\Delta E_{\rm ex}(t)$ is shown. It is calculated by [29]

$$\Delta E_{\rm ex}(t) = \int^{t} \mathbf{E}(t') \cdot \mathbf{J}(t')dt'. \tag{4.20}$$

It is also equal to the difference of the total energies calculated by using the energy density functional. At $I = 10^{11}$ W/cm^2, the excitation energy is substantially small. In the log-scale shown in the inset, the excitation energy is visible during the irradiation of the pulsed electric field and becomes very small after the pulsed electric field ends, since the electronic state goes back almost to the ground state. Contrarily, substantial excitation energy, about 6 eV/atom, is seen at the intensity of $I = 10^{13}$ W/cm^2 due to highly nonlinear interactions.

In Fig. 4.2, we show electron density distributions. In the left panel, the electron density in the ground state is shown in [110] plane. The middle and the right panels show the density change from the ground state at two times, $t = 3.4$ fs when the applied electric field is increasing and $t = 13.5$ fs when the applied electric field

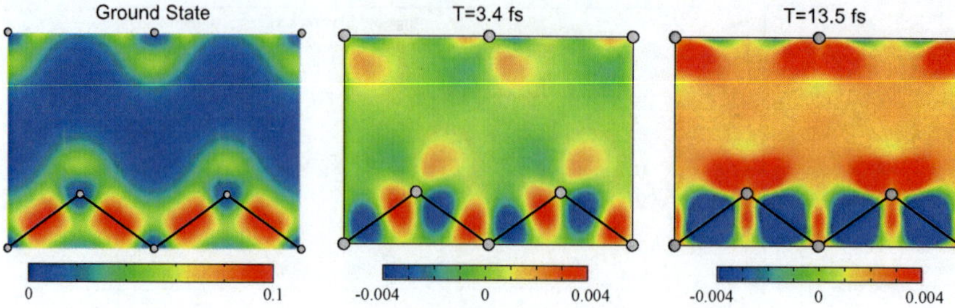

Fig. 4.2 Left panel shows the ground-state electron density. The middle and right panels show the change in the electron density from that in the ground state by a laser pulse. The middle panel corresponds to the time $t = 3.4$ fs and the right panel to the time $t = 13.5$ fs, respectively. In the middle and right panels, the red color indicates the increase in the electron density while the blue color indicates the decrease

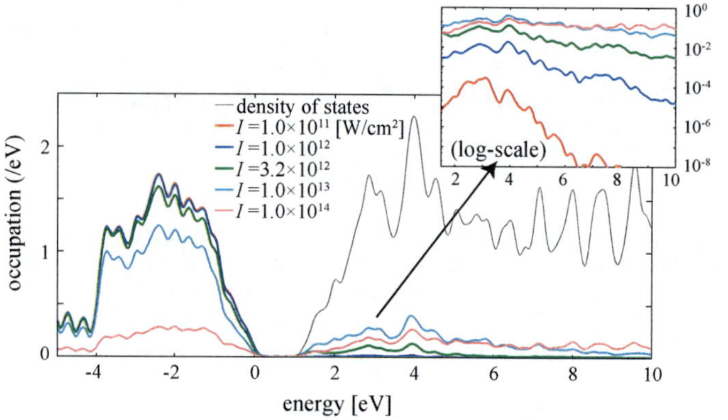

Fig. 4.3 Occupation number distributions in Si after the pulse ends for the pulses of several intensities with $\hbar\omega = 1.55$ eV. Taken from [25] with kind permission of The European Physical Journal (EPJ)

almost ends, respectively. At the time $t = 3.4$ fs, the change of the electron density shows a move of electrons in the Si-Si bond. This corresponds to a virtual excitation of electrons during the irradiation of the pulse. The change after the pulse ends shows a real excitation of electrons by the multiphoton excitation. It shows a move of electrons in the bond region to an interstitial region.

In Fig. 4.3, we show occupation number distributions in Si after the pulse ends for the case of $\hbar\omega = 1.55$ eV. As a function of time t, it is defined by

$$\frac{dn}{d\epsilon} = \sum_{bb'\mathbf{k}} f_{b'\mathbf{k}} \delta(\epsilon - \epsilon^0_{b\mathbf{k}}) |\langle u^0_{b\mathbf{k}} | u_{b'\mathbf{k}}(t) \rangle|^2, \tag{4.21}$$

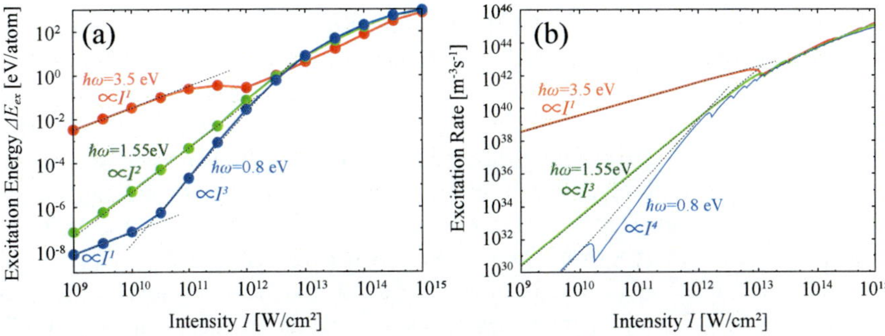

Fig. 4.4 **a** Electronic excitation energies of Si in TDDFT calculation as a function of the intensity of the applied laser pulse with several frequencies. **b** Excitation rates of Si by the Keldysh theory for a continuous wave as a function of the intensity of the applied laser pulse with several chosen photon energies. Taken from [25] with kind permission of The European Physical Journal (EPJ)

where $\epsilon^0_{b\mathbf{k}}$ and $u^0_{b\mathbf{k}}$ is the orbital energy and the Bloch orbital in the ground state. At weaker intensity below $I = 1.0 \times 10^{12}$ W/cm^2, the occupation peak is around 3 eV. At this intensity, it is considered that excitations by two-photon absorption dominate. (See the log-scale plot in the inset.) The occupation at the higher energy region grows as the intensity increases, indicating that transitions within the conduction bands become significant. At the same time, the occupation in the valence bands decreases. It drops drastically when the intensity increases from 10^{13} to 10^{14} W/cm^2, indicating a saturation of the excitation from the valence bands.

In Fig. 4.4a, we summarize our calculated results of the electronic excitation energy, $\Delta E_{\rm ex}(t)$, at a time after the field ends. The excitation energy is shown as a function of the intensity of the applied laser pulse I, for several different frequencies ω. From the figure, we find the following characteristic features.

There is an intersection point in each plot at which the excitation energies of different frequencies merge into the same value at a certain intensity. The intensity of the intersection point is $I = 3.2 \times 10^{12}$ W/cm^2. At intensities above the intersection point, the excitation energy does not depend much on the frequency. The slope becomes small compared with the low intensity region. At intensities below the intersection point, the excitation energies show perturbative behavior and well fit by I^N where N is an integer. The integer number N coincides with the number of photons required to exceed the band gap energy. There are, however, exceptions caused by the spectral width of the short applied pulses. The integer number N is sometimes lower than the number of photons required to across the direct band gap. At $\hbar\omega = 1.55$ eV, three photons are required to across the direct band gap, while the excitation energy is consistent with $N = 2$. At $\hbar\omega = 0.8$ eV, four photons are required, while $N = 3$ fits the excitation energy. At very low intensity region where the excitation energy is below 10^{-6} eV/atom and using low frequency electric field, the energy transfer shows $N = 1$ behavior.

These behaviors are consistent with descriptions using the so-called Keldysh theory of ionization in solids [30]. The Keldysh theory describes strong field ionization phenomena with full inclusion of nonlinearity for an irradiation of a continuum wave field and has been widely used for analyses of experiments. To make a quantitative comparison of our TDDFT results with the Keldysh theory, we show the excitation rate of silicon by the Keldysh theory in Fig. 4.4b where the direct band gap energy value of silicon is set to 3.19 eV, and effective masses of electrons and holes to $1.06m_e$ and $0.59m_e$, respectively. The formula we use here is the Keldysh formula in its simple form. It should be noted that, while the Keldysh theory evaluates only transitions from the valence to conduction bands, the excitation energy in the TDDFT calculation includes the energy transfer in the transitions between the conduction bands after the excitation from the valence to the conduction bands. If we ignore the latter process, the excitation energy $E_{\rm ex}(t = T)$ is related to the excitation rate R_K by $E_{\rm ex}(t = T) = R_K \Delta T(N\hbar\omega)$ where N is the required photon number to exceed the band gap energy and T is the duration of the pulse.

The figure shows that the Keldysh theory reproduces the overall trend of Fig. 4.4a very well, confirming that the fundamental essence of the nonlinear interaction is certainly captured in the Keldysh theory. The intersection point is related to the change from a perturbative to a tunneling regime in which the border of two regimes are divided by the so-called Keldysh parameter, γ. At low intensity region where the Keldysh parameter is much larger than unity, multiphoton absorption is expected. At high intensity region where γ is lower than unity, we expect tunneling and above-barrier ionizations dominate and the frequency of the field becomes less important

4.4 Thin Films

We next show applications of our approach to light-matter interactions in thin films [21]. In very thin films, surface electronic states that are substantially different from bulk states may contribute to the optical response. As the thickness increases, light propagation effects may become important. In this subsection, we will answer the question what is the critical thickness that we need to treat thin films as a 2D/3D material.

4.4.1 Setup of the Problem

We consider an irradiation of a free-standing thin film of thickness d in a vacuum by an ultrashort light pulse of a linearly polarized plane wave at normal incidence. The thin film is assumed to be infinitely periodic in two-dimension (xy-plane) and the thickness d is much less than the wavelength of the incident light. We take the coordinate system as shown in Fig. 4.5. The direction of the light propagation is taken

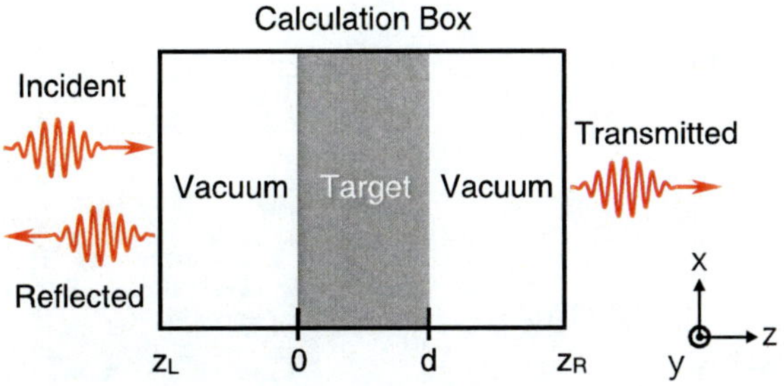

Fig. 4.5 Coordinate system and asymptotic boundary conditions are shown. The direction of the light propagation is along the z axis. The thin material exists in the region, $0 < z < d$. z_L and z_R are the positions sufficiently far apart from the surface of the thin material such that the near field at these positions is negligible. Taken from [21]

along the z-axis, the surfaces of the thin film are set at $z = 0$ and $z = d$ planes. The direction of the polarization of the incident light is taken along the x-axis.

We will describe the problem in three theoretical frames of Maxwell-TDDFT, 3D single-scale, 2D macroscopic, and 3D multiscale. The calculation by 3D Maxwell-TDDFT single-scale method is the most accurate since no macroscopic approximation is introduced, although computationally quite expensive. 2D/3D macroscopic methods include approximations. Comparison of the results by the three methods will provide us with knowledge on what condition the approximate approach may be applicable.

At the beginning of the calculation, the vector potential includes only the incident wave. It locates in the $z < 0$ region and is described as the one-dimensional wave propagation in the z direction,

$$\mathbf{A}(\mathbf{r}, t) = \hat{\mathbf{x}} A^{in} \left(t - \frac{z}{c} \right), \tag{4.22}$$

where $A^{in}(t)$ specifies the time profile of the incident light pulse. At a time long after the interaction, the vector potential consists of the reflected and the transmitted wave, which exist in the spatial region $z < 0$ and $z > d$, respectively. They are expressed as the one-dimensional wave propagating along the z axis as

$$\mathbf{A}(\mathbf{r}, t) = \hat{\mathbf{x}} \mathbf{A}^{re} \left(t + \frac{z}{c} \right) + \hat{\mathbf{x}} A^{tr} \left(t - \frac{z}{c} \right). \tag{4.23}$$

We have made calculations for a thin film of Si of thickness $d = na$ with $n = 1, 3, 10$, and 50 where $a = 5.43$Å is the lattice constant of Si in the cubic unit cell containing eight atoms. Thus the thickness of the films are from $d = a \sim 0.54$ to $d = 50a \sim 27$nm. The atomic positions are not relaxed but are set to those of the bulk silicon. The dangling bonds that appear on surfaces are terminated with hydrogen atoms. For the time profile of the incident pulse, we use the form,

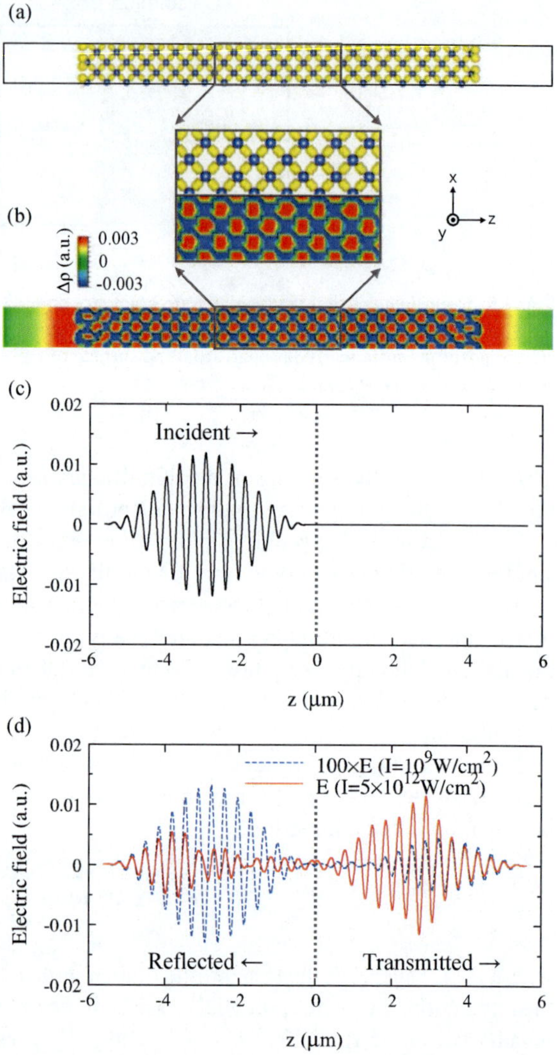

Fig. 4.6 Typical result using the 3D Maxwell-TDDFT single-scale method with the $d = 10a$ thin film. The average light frequency is $\hbar\omega = 3.5$eV. The intensity of the light pulse is $I = 5 \times 10^{12}$ W/cm^2. **a** Electron density in the ground state. The enlarged view of the indicated area is also shown. **b** Change in the electron density from the ground state due to the pulse irradiation. **c** Incident light pulse and the target film ($z = 0$). **d** Reflected and transmitted waves (red-solid line) after the pulse passed through the film ($t = 18.75$ fs). For comparison, the results of weak pulse case ($I = 10^9$ W/cm^2) is also plotted as the blue-dashed line. The latter is scaled up by a factor of 100. Taken from [21]

$$A^{in}(t) = -\frac{E_0}{\omega} \sin\left[\omega\left(t - \frac{T}{2}\right)\right] \sin^2\left(\frac{\pi t}{T}\right), \qquad (0 < t < T), \tag{4.24}$$

where E_0 is the maximum amplitude of the electric field, ω is the average frequency, and T is the pulse duration. The pulse duration is set to $T = 18$ fs and the frequency is changed in the range of $\hbar\omega = 1 \sim 9$ eV. The number of k-points in the 2D reciprocal space is chosen as 4×4. In the calculations shown in this section, we use a simple adiabatic local density approximation [31]. The direct band gap energy then becomes 2.4 eV.

In Fig. 4.6, we show a typical result 3D Maxwell-TDDFT single-scale method. The thickness of the Si film is set to $d = 10a \sim 5.4$nm, and the frequency is $\hbar\omega = 3.5$ eV. This frequency is above the optical gap energy of bulk silicon, and strong absorption of the light pulse is expected. We show the results of two different maximum amplitudes, a sufficiently weak case in which a linear-response theory is applicable and a strong case in which nonlinear light-matter interaction is significant. Figure 4.6a shows the electron number density in the ground state. The high-density region is seen in the bond region. The electric field of the incident pulse is shown in Fig. 4.6c for the case of the strong pulse. The Si film is placed at the $z = 0$ position. Since the thickness of the Si film is much thinner than the wavelength of the incident pulse, the Si film is drawn as a dashed line without thickness in the figure. In Fig. 4.6d, the electric field after the interaction is shown for two cases. The incident pulse is split into transmitted and reflected waves. The blue-dotted curves indicate the results of the weak pulse case, whereas the red-solid curves show the results of the strong pulse case. The electric field of the weak pulse case is rescaled so that it can be directly compared with that of the strong pulse case. At the front part of both transmitted and reflected waves, two curves of the weak and the strong pulses coincide with each other. As the time proceeds, the transmitted wave of the strong pulse case becomes more intense than that of the weak pulse case. On the other hand, the reflected wave of the strong pulse case becomes much smaller than that of the weak pulse case. This fact indicates that the nonlinear light-matter interaction for the strong incident pulse manifests as the field strength becomes strong, in the enhancement in the transmission and absorption, and reduction of the reflection by the Si film. In Fig. 4.6b, the change in electron density from the ground state is shown for the case of strong pulse incidence at the time when the electric field is shown in Fig. 4.6d. The density change is shown after averaging over the y-direction. We can see that the electron density of σ bonds spreads out and fill the spatial region outside the bonds. The absorbed energy is spent to move the binding electrons and to weaken the bonds. These features are the same as that we observed in the bulk Si in Fig. 4.2.

4.4.2 Linear Optical Response

For this problem, we can apply the 2D and 3D macroscopic approaches as well as the microscopic approach. A comparison of the results by three methods will provide us knowledge on what condition the macroscopic approaches will be reliable. In this subsection, we make a comparison for a weak field for which the reflection, transmission, and absorption rates in the 2D/3D macroscopic method can be evaluated using the 2D conductivity introduced in (4.17) and the ordinary 3D dielectric function.

In Fig. 4.7, we show the reflection, transmission, and absorption rates as a function of the frequency for the weak light pulse irradiating on the Si films of various thicknesses. We compare 3D Maxwell-TDDFT single-scale calculations shown by red circles with the rates using 2D and 3D macroscopic descriptions. In the 3D

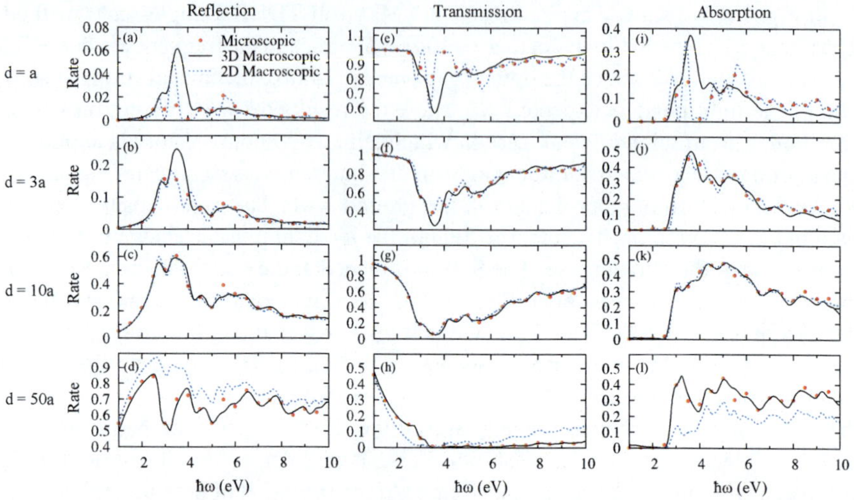

Fig. 4.7 Reflection [**a–d**], transmission [**e–h**], and absorption [**i–l**] rates of the films of thicknesses $d = a$, $3a$, $10a$, and $50a$, respectively, as a function of the frequency $\hbar\omega$ at the time $t = 18.75$ fs. The red circles, black-solid line, and blue-dashed line correspond to the 3D Maxwell-TDDFT single-scale, 3D macroscopic, and 2D macroscopic calculations, respectively. Taken from [21]

Maxwell-TDDFT single-scale calculations, the rates are calculated using $A^{re}(t)$ and $A^{tr}(t)$ immediately after the pulse passed through the film.

Here both effects of the changes in the electronic structure at the surfaces and the multiple reflections at the surfaces appear in the comparison. In the 2D macroscopic description, the electronic structure changes are incorporated through the 2D electric susceptibility, whereas the multiple reflections are not. In the 3D macroscopic description, although the multiple reflections are correctly taken into account, the electronic structure changes are not. We note that the rates of the 3D Maxwell-TDDFT single-scale calculations are for light pulses which include a finite frequency width, whereas those of the 2D and the 3D macroscopic descriptions are for monochromatic waves. As seen from the figure, calculations using the 2D electric susceptibility coincide accurately with the 3D Maxwell-TDDFT single-scale calculations for the films of thickness except for the thickest case $d = 50a$. This indicates that the description using the 2D conductivity of (4.17) is successful for films of thicknesses of 5 nm and less. In the $d = 50a$ case, the absorption using the 2D conductivity is very small compared with the other two calculations. The failure of the 2D description for this case indicates the importance of multiple reflections for that thickness. The description using the 3D electric susceptibility accurately coincides with the 3D Maxwell-TDDFT single-scale calculations for thicker films as may be expected. The difference among the three calculations is very small at $d = 10a \sim 5$ nm. As the thickness decreases, the difference becomes large, and the calculation using the bulk electric susceptibility is poor for the thinnest case of $d = a \sim 0.5$ nm. This is reasonable because the electronic structure of such a thin film is very different from the bulk one.

4.4.3 Nonlinear Optical Response

We next consider nonlinear interaction in thin films when an intense and ultrashort light pulse irradiates normally on the material. We show a systematic view for the nonlinear interaction for the very thin film of $d = a$ thickness using the 3D Maxwell-TDDFT single-scale method. In Fig. 4.8, reflection, transmission, and absorption rates are shown for pulses of various maximum intensities and for several average frequencies. The interaction is in a linear regime below the intensity of 1.0×10^{10} W/cm^2 where the rates do not depend on the intensity. Above 1.0×10^{10} W/cm^2,

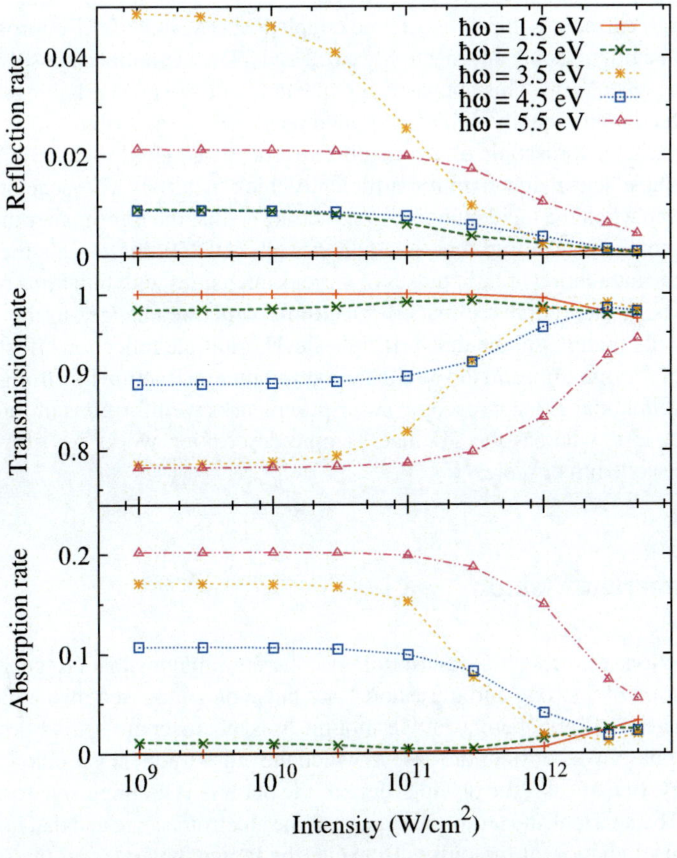

Fig. 4.8 Reflection, transmission, and absorption rates of the film of thickness $d = a$ as a function of the light intensity for several average frequencies. Taken from [21]

the rates change substantially: The reflection and the absorption rates decrease, and the transmission rate increases. This indicates that the intense light pulse tends to transmit through the thin film for light pulses of intensities more than 1.0×10^{10} W/cm^2. Looking in more detail, there is a difference in the intensity at which the nonlinearity starts to appear: For a light pulse of the average frequency 3.5 eV where a strong absorption is seen in the 2D electric susceptibility, the rates start to deviate at the lowest intensity around 1.0×10^{10} W/cm^2. This suggests the saturable absorption as the origin of the nonlinearity: A sizable transition in specific orbitals/k-points causes the saturation of the transition due to the lack of electrons in the occupied orbitals and/or the blocking of the unoccupied orbitals.

4.4.4 Comparison of Different Theoretical Approaches

Summarizing this section, we applied TDDFT approach for the light propagation through extremely thin films. Three coupled Maxwell-TDDFT approaches are applied. In a microscopic approach, Maxwell and TDKS equations are solved using a common spatial grid. This approach is expected to be the most reliable although it costs high computationally. We also applied macroscopic descriptions for two limiting cases: 2D macroscopic electromagnetism for extremely thin materials and 3D macroscopic electromagnetism for sufficiently thick materials. Our schemes can be applied even when the incident field is very strong so that the light-matter interaction is highly nonlinear. We find that the 3D Maxwell-TDDFT single-scale method can describe the interaction of light pulses of various intensities with thin films of various thicknesses. It incorporates effects of electronic structure changes in the thin film from the bulk material in the first-principles level, multiple reflection effects at both surfaces of the thin film, and the nonlinear light-matter interaction for strong incident pulses. We find the 2D macroscopic description works well for films of thicknesses of 5 nm or less, whereas the 3D macroscopic description works well for films of thicknesses of 5 nm or more.

4.5 Coherent Phonon

In this section, we present a description of coherent phonons that are generated by an irradiation of a strong and ultrashort laser pulse on condensed media [32]. They are characterized by coherent lattice motion in a macroscopic spatial area with a common phase. We consider the case in which the pulse frequency is below the band gap energy. In that case, the driving force of the phonon is an impulsive force that is generated by a virtual and temporal change of the electronic state and that only exerts during the irradiation of the pulse. To excite the lattice dynamics by the impulsive force, the pulse duration must be much shorter than the period of the optical phonon.

This nonlinear process of the coherent phonon generation is known as the impulsive stimulated Raman scattering (ISRS) mechanism [33, 34]

To describe the phenomena, we employ the 3D Maxwell + TDDFT + MD multiscale medhod that is capable of describing vast nonlinear optical phenomena involving lattice dynamics [22, 35]. In the method, we solve the following equations simultaneously: For the light propagation, we solve the macroscopic Maxwell equation (4.12). The microscopic electron dynamics are calculated by solving the TDKS equation at each macroscopic grid point **R**, (4.13). At each grid point **R**, we also solve the Newtonian equation (4.3) to describe the lattice motion.

As a material to calculate, we take a diamond thin film. To simulate pump-probe experiments, two ultrashort pulses with the same frequency $\hbar\omega = 1.55$ eV, which is well below the band gap of the diamond, are successively irradiated normally on the surface along the [100] axis. The pulse has a cosine-square envelope with the full time duration of 18 fs, which corresponds to 6.5 fs in full width at half maximum (FWHM). This is shorter than the period of the optical phonon of diamond, which is about 25 fs. First, a strong pump pulse that is linearly polarized along the [011] direction is irradiated. The maximum intensity is set to 2×10^{12} W/cm^2. It generates the coherent optical phonon with atomic displacements in the [100] direction.

Next, a weak probe pulse that is polarized along the [010] direction is used to detect the coherent phonon. The maximum intensity is set to 1×10^{10} W/cm^2 that is sufficiently weak so that there occurs no significant nonlinear effects related to the intensity of the probe pulse and that it does not bring any changes in the lattice dynamics. Accompanying the probe pulse, an impulsively stimulated Raman scattered wave that is polarized along the [001] direction is generated and amplified by the nonlinear interaction between the lattice motion and the probe pulse. Usually, this process is described using the Raman tensor of the diamond. Our simulation automatically includes the effect of solving the dynamical equations. The Raman wave, as well as the propagated probe pulse, appear as transmission signals after they get out of the thin film.

The light propagation in the present setting can be described by the one-dimensional equations. In the calculation, we set the crystalline *abc* axes of the cubic diamond to coincide with the *xyz* axes of the Cartesian coordinates, respectively. Although it is in principle possible to carry out calculations of the pump and the probe pulse propagations in a single calculation, we separate them in two to avoid the complexity coming from the reflection of the pump pulse at the back surface of the thin film. First, we carry out the calculation of the pump pulse, aiming to describe the generation of the coherent optical phonon. At this stage, we carry out the calculation for a sufficiently thick film so no reflections appear at the back surface. In practice, we make a calculation for a diamond thin film of 10 μm thickness and for a time duration of 80 fs. At the end of this calculation, the pump pulse locates in the spatial region 6 μm $< X <$ 10 μm. We extract the lattice dynamics of the spatial region $0 < X <$ 6 μm from this calculation and use it in the next calculation of the probe stage.

For the probe-stage calculation, we consider a thin film of 6 μm thickness. The probe simulation starts with the following initial condition: First, we set the posi-

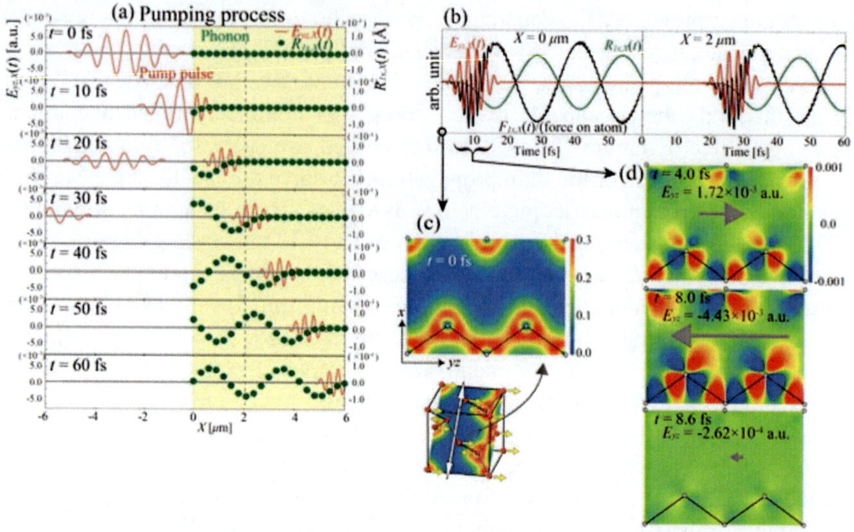

Fig. 4.9 **a** Snapshots of the electric field of the pump pulse (red lines) and the ionic displacement (green filled circles) are shown along the macroscopic grid points of X coordinate. **b** Electric field of the pump pulse (red line), the force acting on the ion (black line), and the ionic displacement (green line) are shown as a function of time at $X = 0$ and 2 μm. **c** The ground-state electron density in the (011) plane of the unit cell. **d** The changes in the electron density from the ground state are shown at $t = 4.0$, 8.0, and 8.6 fs at $X = 0$ μm. Taken from [22]

tions of ions at each macroscopic grid point to the final positions of the pump-stage calculation. We also set the initial velocity of ions. Then, the electronic ground-state calculation is carried out for the shifted ionic positions at each macroscopic point. The probe pulse is then irradiated at the surface.

We first show the calculation for the coherent phonon generation in diamond by the pump pulse. Figure 4.9a shows the propagation of the pump light by red lines and the displacement of the ions at different macroscopic points of X by green filled circles. When the field enters the medium, first the ions at the left edge point of the medium start to move by the light-matter interaction. After that, as the pump pulse propagates through the medium, the harmonic motion of ions is generated in turn at each macroscopic grid point. In Fig. 4.9, a wavelike behavior of the ionic displacement is seen along the X axis. The amplitude of the ionic displacement is rather small, typically 10^{-4} Å. This reflects the fact that there is no resonant energy transfer from the pump pulse to electrons since the pulse frequency is much below the band gap energy.

It is noted that the wavelike behavior is not an ordinary propagation of the lattice wave. The period of the oscillation at each X point is equal to the period of the optical phonon, 25 fs for diamond. However, since the wave is generated by the pump pulse that propagates with the speed of light in the medium, c/n, with the index of refraction n, the front edge of the wave moves with the same speed. This is the order of magnitude larger than the phase velocity of the lattice wave which is determined by the phonon dispersion curve. The amplitude of the oscillation does not attenuate in the medium since the pump pulse is not absorbed during the propagation. Although there is a small energy transfer from the pulsed light to the lattice motion, the energy transfer is very small.

In Fig. 4.9b, the generation processes of the optical phonon at $X = 0$ and $2\ \mu\text{m}$ are shown as a function of time. We first look at the case of $X = 0\ \mu\text{m}$. Immediately after the light electric field arrives at the position, the force that is proportional to the square of the electric field appears. It then generates the ionic displacement corresponding to the coherent phonon. As the ions start to move, the restoring force begins to work. The ionic displacements show a sine shape at times after the center of the pump-pulse envelope (i.e., $t > 9$ fs). These behaviors of the force and the generation process of the coherent phonon are consistent with the picture of the ISRS mechanism. After the pump field passes away, a simple harmonic motion of ions continues without decay. The same generation process is seen at $X = 2\ \mu\text{m}$ in Fig. 4.9b after the pump field arrives at the position. The driving force for ions is generated through the change in the electron density, not from the direct field-ion interaction. In Fig. 4.9c, d electron density in the (011) plane of diamond is shown for the ground state in Fig. 4.9c and changes in density at times $t = 4.0$ fs, 8.0 fs, and 8.6 fs in Fig. 4.9d. In the ground state, the covalent bonds appear between ions as seen in Fig. 4.9c. The pump field induces a shift of bonding electrons along the bond direction toward the direction of the electric field. The larger changes in the density are observed under the stronger field. Although the direction of the shift of electrons changes when the direction of the field becomes opposite, the direction of the force on ions does not change because all bonds in the (011) plane are weakened irrespective of the direction of the electric field. This gives rise to the driving force of ions along the [100] axis during the pump pulse.

We next proceed to the simulations for the propagation of probe pulses. As mentioned previously, we make separate calculations from those of the pump pulse that generates the coherent phonon. The polarization of the probe pulse is set to [010] direction. We express the time delay between the pump and the probe pulses by τ. Two simulations are performed with $\tau = 83.0$ and 89.5 fs. In the calculation of $\tau = 83.0$ fs, the ionic displacement at the surface is maximum when the probe pulse reaches the surface of the medium. In the calculation of $\tau = 89.5$ fs, the ionic displacement at the surface is almost zero when the probe pulse reaches the surface of the medium.

In Fig. 4.10a, the propagation of the probe pulse as well as the generation of the impulsively stimulated Raman wave are shown for the case of the pump-probe time delay of $\tau = 83.0$ fs. As seen from the figure, the probe pulse shown by the red line propagates with the same speed as the wave of the ionic displacement, at the position of the maximum ionic displacement. This is because the lattice motion proceeds with the speed of the pump pulse, c/n, as noted previously, and the probe pulse also propagates with the same speed since we chose the same frequency for the pump and the probe pulses.

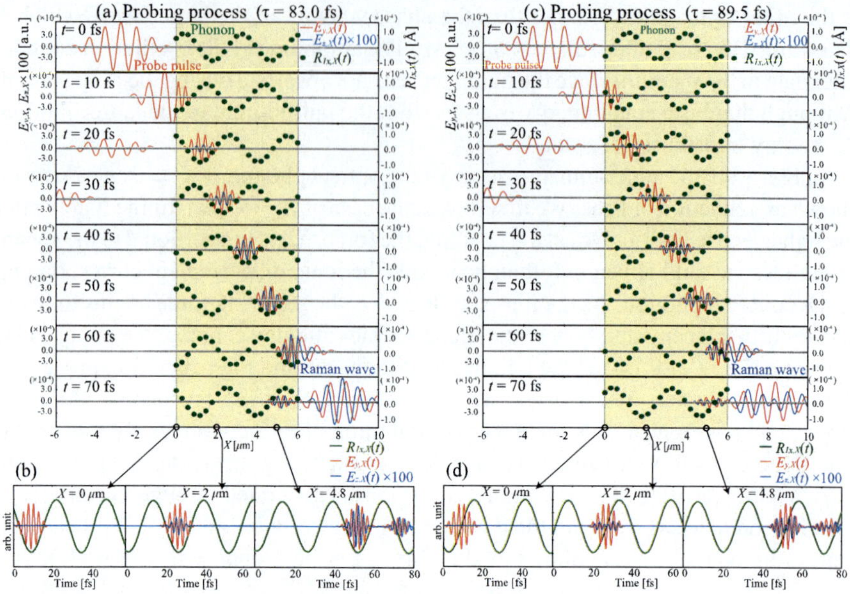

Fig. 4.10 Simulation results of the probe process with $\tau = 83$ fs (**a**, **b**) and with $\tau = 89.5$ fs (**c**, **d**). **a** Snapshots of the electric field of the probe pulse with $\tau = 83$ fs in [010] (red line) and in [001] (blue line) directions, and the ionic displacement (green filled circles) are shown along the macroscopic coordinate X. **b** Plots of the same quantities as a function of time at $X = 0$, 2, and 4.8 μm. The time t is set to 0 when the front edge of the probe pulse reaches the surface of the medium. **c**, **d**. The same as (**a**) and (**b**) except that $\tau = 89.5$ fs. Taken from [22]

During the propagation of the probe pulse, the stimulated Raman wave that is shown by the blue line appears in the [001] direction, perpendicular to both directions of the ionic displacements and the probe polarization. The stimulated Raman wave has the phase shift of $\pi/2$ with respect to the probe pulse and its amplitude increases linearly with the traveled distance. These features are consistent with the standard theoretical description for the stimulated Raman scattering using the property of the Raman tensor.

The probe pulse and the stimulated Raman wave as a function of time are shown at $X = 0$, 2, and 4.8 μm in Fig. 4.10b. It is seen that two waves exist in the timing of the maximum displacement of the ions and have almost the same envelope shape with common center positions. At $X = 4.8$ μm, the probe and the stimulated Raman waves reflected at the back surface of the medium are seen around $t = 75$ fs.

In Fig. 4.10c, the propagation of the probe pulse with the time delay of $\tau = 89.5$ fs is shown. Here, the probe pulse moves with the nodal point of the lattice motion during the propagation in the medium. As is seen from the figure, there appears a clear difference in the pulse shape of the stimulated Raman scattering wave from that of the case of $\tau = 83.0$ fs.

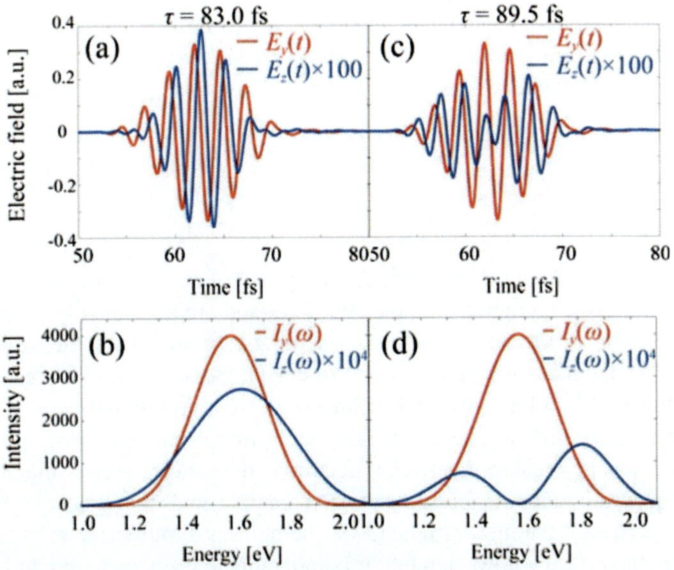

Fig. 4.11 Transmitted probe $[E_y(t)]$ and stimulated Raman waves $[E_z(t)]$ in the right vacuum region for $\tau = 83.0$ fs are shown in (**a**), and their power spectra are shown in (**b**). Those with $\tau = 89.5$ fs are shown in (**c**) and (**d**). Taken from [22]

In Fig. 4.11, we show the pulse shape of the transmitted waves in the time domain for the two cases of different time delays. Figure 4.11a, c are the cases of $\tau = 83.0$ and 89.5 fs, respectively. In the case of $\tau = 83.0$ fs, the pulse shape of the stimulated Raman wave $[E_z(t)]$ is similar to the shape of the probe pulse $[E_y(t)]$ except for the phase difference of $\pi/2$. On the other hand, in the case of $\tau = 89.5$ fs, the pulse shape of the stimulated Raman wave is very different from the shape of the probe pulse. This difference can be understood as originating from the difference of the electric current density that produced the stimulated Raman wave. The electric current density that produces the Raman wave is given by $J_{\text{Raman}}(t) \propto Q(t)E_{\text{probe}}(t)$, where $E_{\text{probe}}(t)$ is the electric field of the probe pulse and $Q(t)$ is the phonon amplitude that is expressed by the linear combination of the ionic displacements. When the probe pulse enters the medium at the maximum of the phonon amplitude, $Q(t)$ may be regarded as roughly a constant and $J_{\text{Raman}}(t)$ has a similar time profile to that of $E_{\text{probe}}(t)$ since the half period of the lattice motion is assumed to be longer than the duration of the probe pulse. However, when the probe pulse moves with the nodal position of the lattice motion, the phonon amplitude may be approximated by a linear function of time. Then, we have $J_{\text{Raman}}(t) \propto tE_{\text{probe}}(t)$, and the lattice motion produces one extra node to the electric current density. This explains the shape change of the stimulated Raman wave shown in Fig. 4.11c.

We show the power spectra of the stimulated Raman waves in Fig. 4.11b, d. Reflecting the different profiles in time domain, they show distinct structures: the double-peak structure appears in the power spectrum of the stimulated Raman wave when it propagates at the nodal point of the lattice motion. We note that such a double-peak structure is indeed related to the recent pump-probe measurement of the coherent phonon in diamond and other insulators, as reported in [36, 37].

4.6 Metasurfaces

Metasurfaces are artificial sheet materials composed of nano-objects whose sizes are smaller than the wavelength of the incident light and are periodically placed on the planar sheet. They have been attracting substantial attention due to their quite useful optical properties that may be beneficial to design various optoelectronic devices. We will illustrate two kinds of applications of our computational approach for the optical response of metasurfaces: One is the microscopic calculation of a plasmonic metasurface where a distance between metallic nanoparticles are so small that there appears a quantum tunneling current between nanoparticles that dramatically affects optical properties of the metasurface [23]. The other is a multiscale calculation of a dielectric metasurface whose diameter is comparable to the wavelength of light [24].

4.6.1 Plasmonic Metasurface with a Sub-nm Gap

In the past decade, tremendous attention has been paid to subnanometer gap plasmonics in isolated systems with metallic nanodimers constituting a typical paradigm [38, 39]. In brief, the research findings were based on interesting observations in which substantial differences existed between descriptions using classical and quantum theories. In the classical description, in which the surface of nanoparticles are modeled as discontinuous boundaries, the solutions of Maxwell's equations yield an extremely enhanced and confined electromagnetic field at the gap as the nanoparticles come closer to each other. By contrast, in quantum description with continuous boundaries expressed using realistic electron density distributions, the charge transfer processes across the gap that were caused by electron tunneling suppressed the field enhancements, especially at the small gap distances (≤ 0.4 nm). This suppression in the gap plasmonics has been confirmed by experiments [40].

Optical properties of metasurfaces are usually explored for those composed of nano-objects whose sizes and gap distances are at the wavelength or subwavelength scales. Recently, however, experimental studies have reported on metasurfaces with gap distances as small as a subnanometer. For example, the optical properties of a plasmonic metasurface with subnanometer gaps made of gold nanospheres coated with tunable alkanethiol ligand shells were reported [41] in which the interparticle distance was controlled from 0.45 to 2.8 nm. It has been found that owing to the significantly intense optical confinement at the gaps, the metasurface yielded a high-refractive index that could not be achieved in naturally occurring transpar-

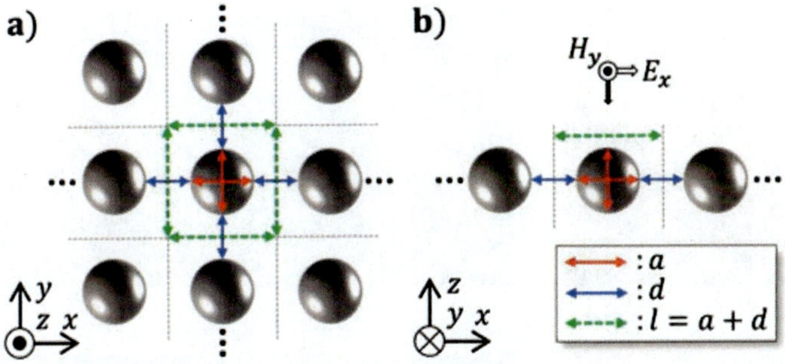

Fig. 4.12 Schematic illustration of the metasurface consisting of periodically arrayed nanospheres. Reprinted with permission from [23]. Copyright 2019 American Chemical Society

ent materials. In view of the broad applicability of metasurfaces, it is particularly concerning how the quantum effects appear in them.

We thus carried out calculations of quantum plasmonic metasurfaces composed of metallic nanospheres with subnanometer gaps using a fully quantum mechanical method, TDDFT with 2D macroscopic approximation [23]. In Fig. 4.12, the structure of the metasurface to be studied is illustrated. Spherical metallic nanoparticles of a diameter a of 3.1nm is periodically arrayed with the gap distance d. To model the nanoparticle, we use a jellium model in which ions are treated as a uniform charge distribution of diameter a. We apply the 2D macroscopic approximation with linear response: The optical response is characterized by the 2D conductivity $\sigma_{2D}(\omega)$ as discussed in (4.17) and below. In the TDDFT calculation, we solve the TDKS equation with a periodic boundary condition in xy-direction and an isolated boundary condition in z-direction. We also solve the same problem within classical electromagnetism assuming a Drude model for the medium of the nanoparticle.

Figure 4.13 shows the calculated spectral distributions of the transmission (T), the reflection (R), and the absorption rates (A) of the plasmonic metasurfaces for gap distances ranging from $d = -0.2 \sim 2$ nm. Figure 4.13a–c shows results based on the use of the classical theory, while Fig. 4.13d–f shows the results estimated based on the TDDFT with the use of the jellium model.

For $d \geq 0.4$nm, the quantum and classical descriptions generate almost similar results that are characterized by a single peak derived from the bonding dipole plasmon (BDP) mode. These features match well with those reported in the previous investigation [41] whereby the plasmonic gap distance was explored to be as small as 0.45 nm in the measurements and simulations according to the classical theory. When $d \leq 0.4$nm, the bonding octopolar plasmon (BOP) mode emerges, and is clearly visible in the classical description at a higher frequency region. By decreasing the distance to $d = 0$nm or less, the void plasmon (VP) and its higher-order plasmon (VP') modes appear when the classical theory is used.

We now look into the electron transport effects at small d regions. As observed in Fig. 4.13d–f, the BDP mode excited on the metasurface survives within a distance

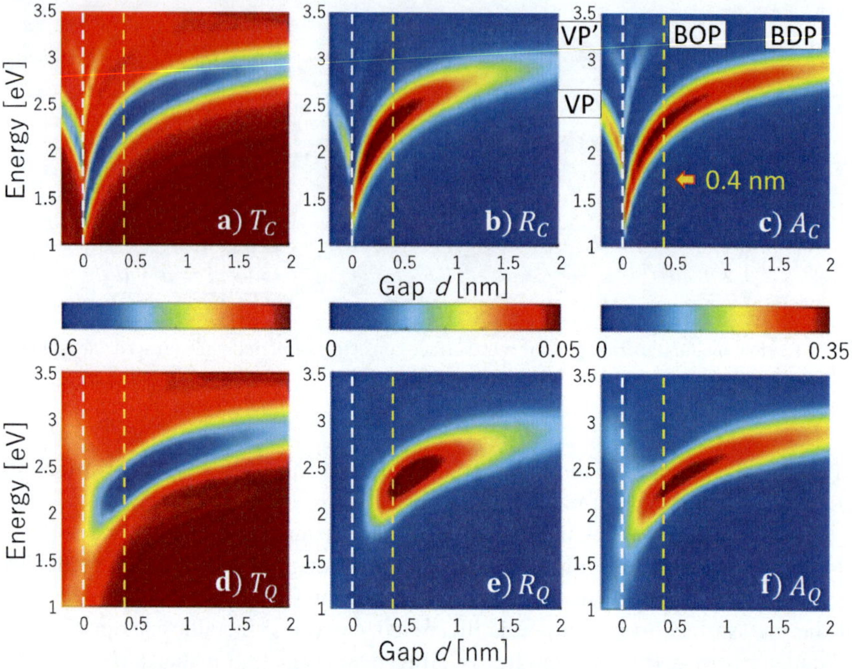

Fig. 4.13 Spectral distribution of the transmission (T), reflection (R), and absorption (A) rates of the metasurfaces calculated by the classical theory (subscript C) and by the TDDFT (subscript Q). All distributions are shown as functions of the gap distance d and the optical frequency. Dashed lines indicate $d = 0$ and 0.4nm. Reprinted with permission from [23]. Copyright 2019 American Chemical Society

as small as $d \sim 0.2$nm. It indicates that the studied metasurface maintains a strong optical confinement and enhancement in the gap distance as small as $d \sim 0.2$nm, which is much smaller than the critical distance $d \sim 0.4$nm in the case of nanodimer. Furthermore, although the classical theory yields similar spectral distributions among T, R, and A, those calculated by the TDDFT yield a difference at $d \leq 0.2$nm, R related to the BDP vanishes immediately, while T and A remain and are extensively spread in frequency, representing washed-out plasmonic features of the constituent nanospheres in the metasurface. In particular at $d = 0$nm, R is less than 0.5%, while A is maintained at 10% in the visible frequency range. This indicates that they are suitable for an optical absorber [42].

To elucidate the mechanism that produces the differences between the classical and quantum descriptions, we visualized the spatial distributions of the electric current density, $j(r, \omega)$, calculated by the TDDFT. For an electric field with frequency ω, $E(t) = E_0 \cos \omega t$, the electric current density is given by $j(r, t) = \mathrm{Re}[j(r, \omega)] \cos \omega t - \mathrm{Im}[j(r, \omega)] \sin \omega t$. We refer to $\mathrm{Re}[j(r, \omega)]$ as the conductive current because it has the same phase relation to the applied electric field. Additionally, we refer to $\mathrm{Im}[j(r, \omega)]$ as the dielectric because it has a phase difference of

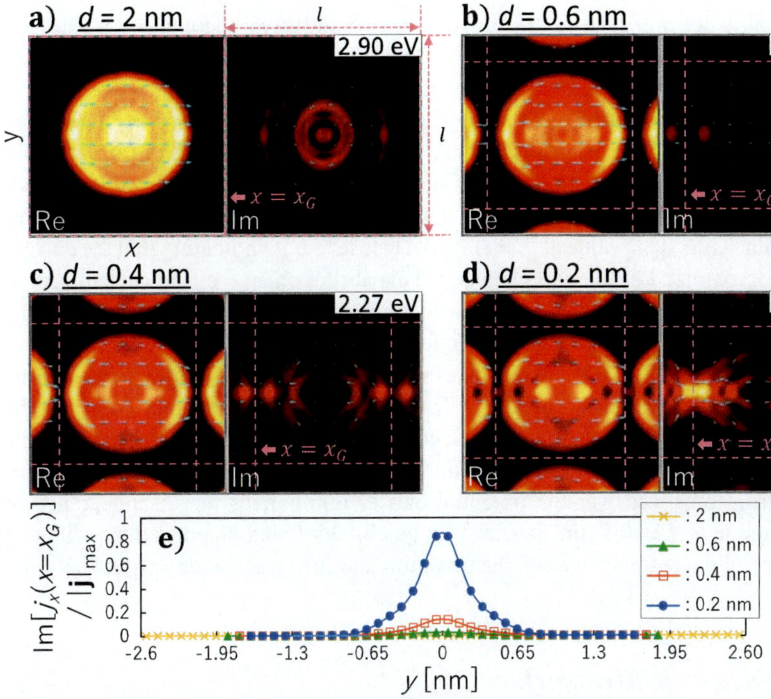

Fig. 4.14 **a–d** Spatial distributions of electric current density on the xy-plane ($z = 0$) at the resonant frequency of the bonding dipolar plasmon (BDP) mode (shown in the top-right white boxes). The maximum intensity is normalized in each figure, showing $j(x, y, z = 0, \omega = \omega_{BDP})/|j|_{\max}$. Real and imaginary parts are shown in the left and right panels, respectively, for $d = 2$, 0.6, 0.4, and 0.2 nm. The light blue arrows indicate their vector distributions whose lengths are associated with their magnitudes. Dashed pink lines indicate the unit cell with a period l. The electron transport takes place through the $x = x_G$ plane. **e** The imaginary parts of the currents are displayed as a function of y at the plane $x = x_G$. Reprinted with permission from [23]. Copyright 2019 American Chemical Society

$\pi/2$. Figure 4.14a–d shows the electric current density in the xy-plane ($z = 0$) at the resonant frequencies of the BDP mode, ω_{BDP}. The maximum value is normalized in each figure, showing $j(x, y, z = 0, \omega = \omega_{\mathrm{BDP}})/|j|_{\max}$. The real and imaginary parts of the current are shown in the left and right panels, respectively, for $d = 2$, 0.6, 0.4, and 0.2 nm. The light blue arrows indicate the vector distributions of the current, whereby their lengths are associated with the magnitudes. The dashed pink lines indicate the unit cell with a period l. At $d = 2$ nm, Re[j] is almost spatially uniform inside the sphere, as observed in Fig. 4.14a. As d decreases, the currents tend to be localized at both ends of the sphere. This localization, which causes the red shift of the BDP mode, as observed in Fig. 4.13, is caused by the attractions among the positive and negative charges that appear at both ends of the neighboring nanospheres. This behavior is also observed in the classical description. By contrast, as shown in Fig. 4.14d, at $d = 0.2$ nm, the current distribution in the TDDFT shows a drastic change that is not observed in the classical theory. In Re[j], there appears a hole in the electric current distribution at both ends of the sphere along the $y = 0$ axis. In Im[j], a clear opening of the current owing to the electron transport is observed in the $x = x_G$ plane along the y axis. To elucidate the electron transport clearly, we plot the current distribution Im[j] against y in the $x = x_G$ plane in Fig. 4.14e using the gap distances of Fig. 4.14a–d. This shows that the electron transport increases gradually as d decreases up to around 0.4 nm, and then increases abruptly toward $d = 0.2$ nm. The $|j|_{max}$ value at $d = 0.2$ nm is close to 0.9, indicating that the current is almost maximized at the gap plane x_G. This abrupt change of the current can be rationalized by the change of the relation between the potential barrier height and the energy of the highest occupied orbital. In fact, we have found that they coincide with each other at around $d = 0.3$ nm. Therefore, the electron transport process can be classified into two situations: (a) At $d \geq 0.3$ nm, electron transport takes place by the quantum tunneling; (b) at $d \leq 0.3$ nm, electron transport is dominantly caused by the overbarrier process so that a large amount of electrons can move through the gaps. The electron transport over the potential barrier rationalizes the disappearance of the BDP mode at $d = 0.2$ nm, as shown in Fig. 4.13d–f, and displays the qualitative differences in the results between the quantum and the classical descriptions.

4.6.2 Dielectric Metasurface

We next consider nonlinear optical responses of metasurfaces composed of dielectric nano-objects [24]. Specifically, we consider a simple case of Si nanospheres with a radius of 250 nm. For such nano-objects, optical responses can be described using macroscopic electromagnetism. To describe extremely nonlinear responses caused by intense and ultrashort laser pulses, it is necessary to make a coupled calculation of the macroscopic light propagation with the electronic dynamics in the microscopic

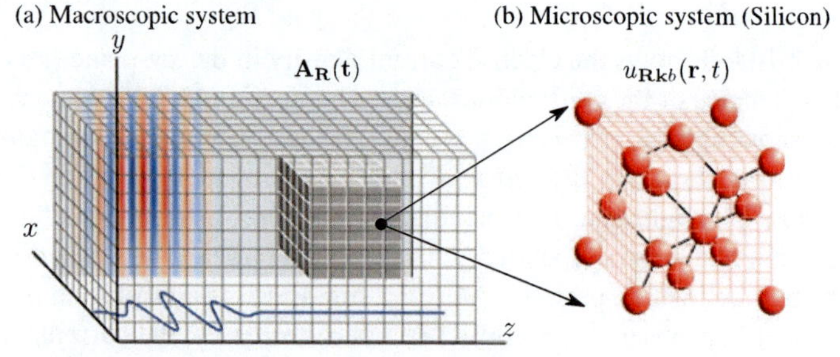

Fig. 4.15 Schematic illustration of 3D Maxwell-TDDFT multiscale simulation. **a** Macroscopic grid system to solve macroscopic Maxwell equation. **b** Microscopic grid system to calculate electron dynamics by solving TDKS equation. Taken from [24] with kind permission of The European Physical Journal (EPJ)

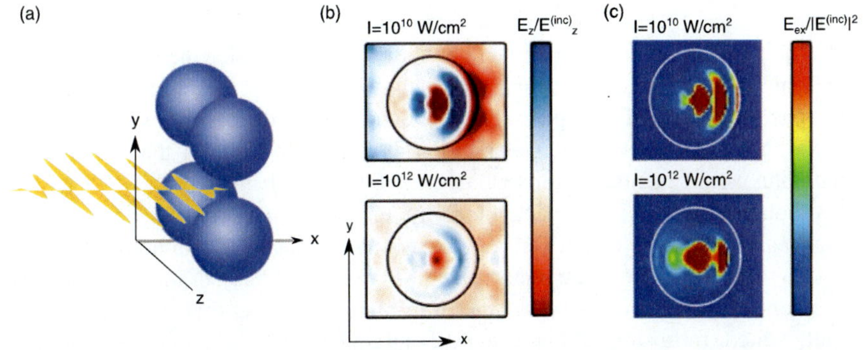

Fig. 4.16 **a** Schematic illustration of the considered system; a linearly polarized pulsed light irradiate normally on silicon nanospheres that are periodically placed in yz-plane. **b** Electric field E_z and **c** electron excitation energy E_{ex}, at $t = 18.4$ fs for two incident intensities ($I = 10^{10}$and 10^{12} W/cm^2). Taken from [24] with kind permission of The European Physical Journal (EPJ)

scale. Therefore, we utilize the 3D Maxwell-TDDFT multiscale method solving the macroscopic Maxwell equation (4.12) and the microscopic TDKS equation (4.13), simultaneously. The scheme to solve the Maxwell and TDKS equations simultaneously is illustrated in Fig. 4.15.

As illustrated in Fig. 4.16a, we place silicon nanospheres in yz-plane periodically. A linearly polarized pulsed light is applied normally to the plane. We set the frequency $\hbar\omega = 1.55$ eV and consider cases of two different maximum intensities. The silicon nanosphere is expressed by 32,752 macroscopic grid points, and electron dynamics calculations are carried out at each point solving the TDKS equation. Since it requires very large computational resources, we have made highly efficient parallelizatiion in our code and used a parallel supercomputer for the calculation [43].

Since the direct band gap of Si is much higher than the frequency of the pulsed light, electronic excitations only take place through nonlinear (mainly two-photon absorption) processes in the present calculation. We expect strong electronic excitations for the case of strong ($I = 10^{12}$ W/cm^2) pulsed light, while negligible excitations for the weak ($I = 10^{10}$ W/cm^2) case. This suggests an immediate change in the optical response of Si nanospheres from dielectric to metallic for the case of the strong pulse.

We show snapshots of the calculated electric field around the Si nanosphere in Fig. 4.16b, and of the electronic excitation energy inside the nanosphere in Fig. 4.16c. The figure is colored so that it looks the same if the optical response is linear. In Fig. 4.16b, a generation of a confined field like a whispering-gallery wave is observed inside the Si nanosphere. A focus is seen around the center of the sphere. The electromagnetic fields look weakened for stronger pulsed light. This is because the nonlinear absorption of the strong light makes the lifetime of the confined light modes much shorter. In Fig. 4.16c, intense electronic excitations are seen inside the sphere. The intensity of the excitation scales approximately linear to the field intensity, indicating that this is not due to the two-photon absorption. Since the figure is drawn at a time where a part of the light field is confined inside the nanosphere, the electronic excitation energy is mainly associated with the polarization dynamics of electrons. In the case of stronger pulsed light, the electronic excitations are a mixture of two-photon absorption and polarization dynamics. We expect that a long-time simulation would eventually separate two excitation mechanisms.

Although this kind of calculations require extremely large computational resources, the multiscale approach will be useful to understand and predict the phenomena that take place in the metasurface where the interplay of the near field and the nonlinear electron dynamics comes into play. We expect this approach will be ideally suited, for example, for recent measurements of high harmonic generation from metasurfaces [44].

4.7 Summary

We have been developing an ab initio computational approach that should be quite useful in the current frontiers of nanophotonics. Our theoretical scheme is based on time-dependent density functional theory that describes microscopic dynamics of electrons and ions in matters. We developed a few schemes that combine microscopic TDDFT calculations for electrons and ions with the time evolution of electromagnetic fields of light. By this way, it becomes possible to describe various light-matter interactions in nano-objects in the first-principles level.

In this article, we mainly presented applications in simple materials such as silicon, diamond, and a jellium model for metals. However, the approach will be readily applicable to any materials for which density functional approaches are applicable. The methods presented in this article are implemented in an open-source code SALMON [28] (Scalable Ab initio Light-Matter simulator for Optics and Nanoscience) that is developed and maintained in our group, and is downloadable from our website [45]. We hope the code, as well as the methods presented here, will be widely applied in many frontiers of nanophotonics.

Acknowledgements This research was supported by JST-CREST under grant number JP-MJCR16N5, and by MEXT, Japan as a priority issue theme 7 to be tackled by using Post-K Computer, and by JSPS KAKENHI, Japan Grant Numbers 20H02649, 20K15194, 19K05364.

References

1. C.A. Ullrich, *Time-Dependent Density-Functional Theory Concepts and Applications* (Oxford University Press, 2012)
2. E. Runge, E.K.U. Gross, Phys. Rev. Lett. **52**, 997 (1984)
3. K. Cho, Optical Response of Nanostructures: Microscopic Nonlocal Theory. Springer Series in Solid-State Sciences (2003)

4. T. Brabec, F. Krausz, Rev. Mod. Phys. **72**, 545 (2000)
5. K. Yabana, G.F. Bertsch, Phys. Rev. B **54**, 4484 (1996)
6. G.F. Bertsch, J.-I. Iwata, A. Rubio, K. Yabana, Phys. Rev. B **62**, 7998 (2000)
7. K. Yabana, G.F. Bertsch, Int. J. Quantum Chem. **75**, 55 (1999)
8. T. Nakatsukasa, K. Yabana, J. Chem. Phys. **114**, 2550 (2001)
9. J.-I. Iwata, K. Yabana, G.F. Bertsch, J. Chem. Phys. **115**, 8773 (2001)
10. K. Nobusada, K. Yabana, Phys. Rev. A **70**, 043411 (2004)
11. Y. Kawashita, T. Nakatsukasa, K. Yabana, J. Phys. Cond. Matter **21**, 064222 (2009)
12. M. Uemoto, Y. Kuwabara, S.A. Sato, K. Yabana, J. Chem. Phys. **150**, 094101 (2019)
13. Y. Shinohara, K. Yabana, Y. Kawashita, J.-I. Iwata, T. Otobe, G.F. Bertsch, Phys. Rev. B **82**, 155110 (2010)
14. T. Otobe, M. Yamagiwa, J.-I. Iwata, K. Yabana, T. Nakatsukasa, G.F. Bertsch, Phys. Rev. B **77**, 165104 (2008)
15. T. Otobe, Phys. Rev. B **94**, 235152 (2016)
16. G. Wachter, C. Lemell, J. Burgdörfer, S.A. Sato, X.-M. Tong, K. Yabana, Phys. Rev. Lett. **113**, 087401 (2014)
17. K. Yabana, T. Sugiyama, Y. Shinohara, T. Otobe, G.F. Bertsch, Phys. Rev. B **85**, 045134 (2012)
18. M. Lucchini, S.A. Sato, A. Ludwig, J. Herrmann, M. Volkov, L. Kasmi, Y. Shinohara, K. Yabana, L. Gallmann, U. Keller, Science **353**, 916 (2016)
19. A. Sommer, E.M. Bothschafter, S.A. Sato, C. Jakubeit, T. Latka, O. Razskazovskaya, H. Fattahi, M. Jobst, W. Schweinberger, V. Shirvanyan, V.S. Yakovlev, R. Kienberger, K. Yabana, N. Karpowicz, M. Schultze, F. Krausz, Nature **534**, 86 (2016)
20. S.A. Sato, K. Yabana, Y. Shinohara, T. Otobe, K.-M. Lee, G.F. Bertsch, Phys. Rev. B **92**, 205413 (2015)
21. S. Yamada, M. Noda, K. Nobusada, K. Yabana, Phys. Rev. B **98**, 245147 (2018)
22. A. Yamada, K. Yabana, Phys. Rev. B **99**, 245103 (2019)
23. T. Takeuchi, M. Noda, K. Yabana, ACS Photon. **6**, 2517 (2019)
24. M. Uemoto, K. Yabana, S.A. Sato, Y. Hirokawa, T. Boku, EPJ Web of Conference **205**, 04023 (2019)
25. A. Yamada, K. Yabana, Euro. Phys. J. D **73**, 87 (2019)
26. N. Troullier, J.L Martins: Phys. Rev. B **43**, 1993 (1991)
27. F. Tran, P. Blaha, Phys. Rev. Lett. **102**, 226401 (2009)
28. M. Noda, S.A. Sato, Y. Hirokawa, M. Uemoto, T. Takeuchi, S. Yamada, A. Yamada, Y. Shinohara, M. Yamaguchi, K. Iida, I. Floss, T. Otobe, K.-M. Lee, K. Ishimura, T. Boku, G.F. Bertsch, K. Nobusada, K. Yabana, Comput. Phys. Commun. **235**, 356 (2019)
29. S.A. Sato, Y. Taniguchi, Y. Shinohara, K. Yabana, J. Chem. Phys. **143**, 224116 (2015)
30. L.V. Keldysh, Sov. Phys. JETP **20**, 1307 (1965)
31. J.P. Perdew, A. Zunger, Phys. Rev. B **23**, 5048 (1981)
32. R. Merlin, Solid State Comm. **102**, 207 (1997)
33. L. Dhar, J.A. Rogers, K.A. Nelson, Chem. Rev. **94**, 157 (1994)
34. M. Hase, M. Kitajima, J. Phys. Cond. Matter **22**, 073201 (2010)
35. A. Yamada, K. Yabana, Phys. Rev. B **101**, 214313 (2020)
36. K.G. Nakamura, K. Ohya, H. Takahashi, T. Tsuruta, H. Sasaki, S. Uozumi, K. Norimatsu, M. Kitajima, Y. Shikano, Y. Kayanuma, Phys. Rev. B **94**, 024303 (2016)
37. K. Mizoguchi, R. Morishita, and G. Oohata: Phys. Rev. Lett. textbf110, 077402 (2013)
38. J. Zuloaga, E. Prodan, P. Nordlander, Nano Lett. **9**, 887 (2009)
39. G. Aguirregabiria, D.C. Marinica, R. Esteban, A.K. Kazansky, J. Aizpurua, A.G. Borisov, Phys. Rev. B **97**, 115430 (2018)
40. J.A. Scholl, A. García-Etxarri, A.L. Koh, J.A. Dionne: Nano Lett. **13**, 564 (2013)
41. D. Doyle, N. Charipar, C. Argyropoulos, S.A. Trammell, R. Nita, J. Naciri, A. Piqu, J.B. Herzog, J. Fontana, ACS Photonics **5**, 1012 (2018)
42. K. Aydin, V.E. Ferry, R.M. Briggs, H.A. Atwater, Nat. Commun. **2**, 517 (2011)
43. Y. Hirokawa, T. Boku, M. Uemoto, S.A. Sato, K. Yabana: Proc. ISC High Performance 2018, Lecture Notes in Computer Science, Vol. **10876**, 202 (2018)
44. H. Liu, C. guo, G. Vampa, J.L. Zhang, T. Sarmiento, M. Xiao, P.H. Bucksbaum, J. Vuckovic, S. Fan, D.A. Reis: Nature Phys. **14**, 1006 (2018)
45. Web site of SALMON: https://salmon-tddft.jp

Chapter 5
Optical Spectroscopy of Individual Single-Walled Carbon Nanotubes

Kaihui Liu, Yuichiro K. Kato, and Shigeo Maruyama

Abstract Carbon nanotubes represent one of the most unique one-dimensional material in the field of nanotechnology. Because of their unique one-dimensionality combined with structure-related electronic and optical properties, single-walled carbon nanotubes exhibit rich physics and attractive applications. Through the overview of the typical spectroscopy such as absorption/Rayleigh scattering, photoluminescence spectroscopy and resonance Raman spectroscopy, we summarize recent efforts in probing and modulating excitonic features of individual single-walled carbon nanotubes. Finally, we will also discuss the recent advances in the study of single-walled carbon nanotubes for future application in photonics.

5.1 Introduction

The last thirty years have witnessed intensive research on carbon nanotubes since the first paper published by Sumio Iijima [1]. The electrical and optical properties of a nanotube depend sensitively on the precise tube structure, which can vary from semiconducting to metallic with the slightest change of chiral index. This richness and diversity of carbon nanotubes make them appealing for a variety of very different applications. Although applications based on semiconducting properties of the

K. Liu
State Key Laboratory for Mesoscopic Physics, Frontiers Science Center for Nano-optoelectronics, School of Physics, Peking University, 209 Chengfu Rd, Haidian District, Beijing, China
e-mail: khliu@pku.edn.cn

Y. K. Kato
Nanoscale Quantum Photonics Laboratory, RIKEN Cluster for Pioneering Research and Quantum Optoelectronics Research Team, RIKEN Center for Advanced Photonics, 2-1 Hirosawa, Wako, Saitama, Japan
e-mail: yuichiro.kato@riken.jp

S. Maruyama (✉)
Department of Mechanical Engineeing, School of Engineering, The University of Tokyo, 7-3-1 Hongo, Bunkyo, Tokyo, Japan
e-mail: maruyama@photon.t.u-tokyo.ac.jp

T. Yatsui (ed.), *Progress in Nanophotonics 6*, Nano-Optics and Nanophotonics,
https://doi.org/10.1007/978-3-030-71516-8_5

single-walled carbon nanotube (SWCNT) [2] are still on the way, research on their conductive, mechanical and thermal properties have developed to a certain level.

Especially, spectroscopic characterization of individual SWCNTs have rapidly developed since the resonant Raman spectroscopy of SWCNTs was first measured in 1997 [3]. The strong resonant features of Raman spectra were determined [4], and radial breathing mode (RBM) peak was first used for chirality assignment of individual SWCNTs on substrates in 2002 [5]. Based on this, Kataura plot relating the energy of the band gap and the diameter of carbon nanotubes has been established.

Another spectroscopy technique, photoluminescence (PL) spectroscopy, was also introduced for micelle-suspended SWCNTs in 2002 [6]. It quickly become an essential tool for the determination of chirality distribution of semiconducting carbon nanotubes, in particular for the development of the chirality separation techniques by density-gradient ultracentrifuge (DGU) [7], DNA-based dispersion [8], and gel-chromatography [9]. Soon after, the strong PL from individual SWCNT were studied [10]. Recently, combining with silicon based photonic crystals [11] and defect-based color-centers, single-photon emission of individual SWCNTs were also detected.

In addition, SWCNTs can be probed optically by elastic light scattering. This effect forms the basis of a technique, termed Rayleigh scattering spectroscopy, for the study of individual nanotubes [12]. Spectroscopic information on the electronic transitions of both semiconducting and metallic nanotubes obtained by measuring the elastic scattering cross section as a function of photon energy is another breakthrough in the optics research of SWCNTs [13].

Until recently, optical spectroscopy has developed as a metrology technique (characterization of samples) rather than a unique application for photonic devices. Since chirality specific growth of SWCNT is still a challenge, chirality separation techniques of micelle-dispersed nanotubes are developed along with absorption, PL, and Raman spectroscopy. In addition, people have enjoyed the rich and peculiar physics of the 1D nanomaterial such as strong environmental effect and strong Coulomb interactions which influence spectroscopic features of SWCNT since the typical size of excitons is much larger than the diameter of SWCNTs. In this review, to concentrate on the pristine optical features of SWCNTs, we mainly focus on spectroscopy and photonics of *individual* SWCNTs. After a quick summary of absorption, PL, and Raman optical techniques, we will discuss recent advances in double-walled carbon nanotubes, hetero-nanotubes, Luttinger-liquid plasmons and coupling to silicon photonic devices. This review presents an overview of optical spectroscopy of SWCNTs for future application in photonics.

5.2 Optical Spectroscopy of Single-Walled Carbon Nanotubes

A SWCNT can be viewed as a hollow cylinder formed by rolling up a graphene sheet, and its geometrical structures can be uniquely defined by its chiral index (n, m) [14, 15]. In the zone-folding approach, without considering the many-body coulomb interactions, the electronic structure of carbon nanotubes can be obtained by cutting the Dirac cone of graphene with a series of parallel **k** lines [14, 15] (Fig. 5.1). For those nanotubes with chiral index mod $(n - m, 3) = 0$, one of the parallel **k**-lines passes through the Dirac point in graphene, resulting in metallic properties. Other nanotubes with mod $(n - m, 3) = 1$ and 2 are semiconducting ones; named mod1 and mod2, respectively. It's worth noting that some people categorize semiconducting SWCNT based on mod $(2n + m, 3) = 1$ and 2 as 'type-I' ('SI' or 'mod 2') and 'type-II' ('SII' or 'mod 1'), respectively.

In the Brillouin zone of nanotubes, each parallel **k**-line describes one pair of conduction and valence sub-bands. Given that those parabolic sub-bands provide a series of singularities in the electronic density of states, transitions between sub-band pairs lead to strong optical resonances, which are traditionally labeled as S_{ii} for semiconducting and M_{ii} for metallic nanotubes, where i is the sub-band index (Fig. 5.2).

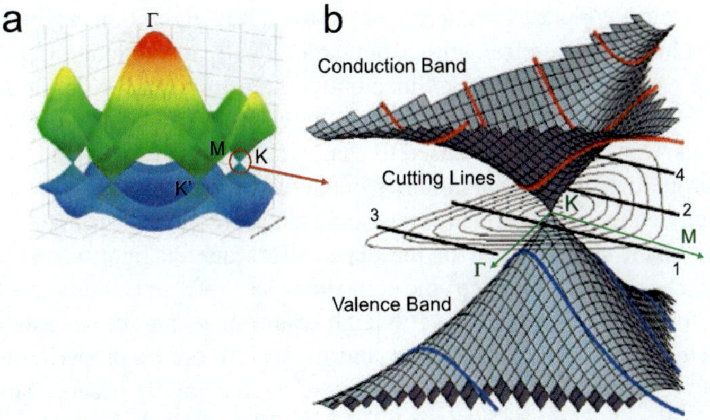

Fig. 5.1 Electronic structure of single-walled carbon nanotubes. The cutting lines and dispersion are drawn for (7,5) SWCNT. Part of the drawings are courtesy of Prof. Yuhei Miyauchi at Kyoto University and Mr. Kaoru Hisama at UTokyo

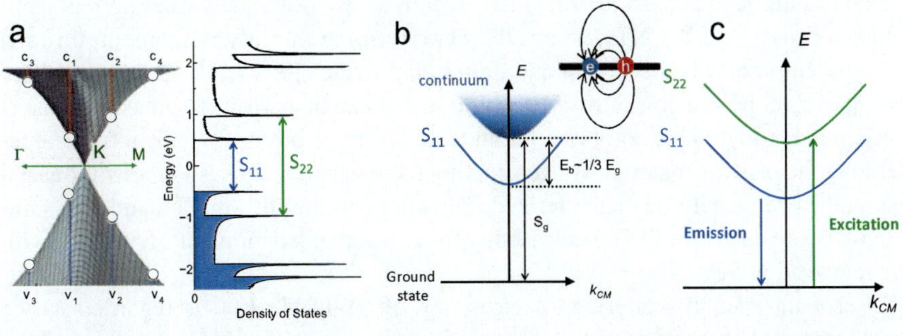

Fig. 5.2 Density of states and excitonic pictures. **a** Dispersion and density of states for (7,5). **b** Exciton binding energy. **c** Typical PL process

The origin of those distinct optical resonances in carbon nanotubes not only stem from the van Hove singularities, but also from strong excitonic effects. In the one-dimensional (1D) SWCNTs, an electron-hole pair can form a bound state known as exciton. Optical transitions are more accurately described as the generation and recombination of excitons [16]. The traditional labels S_{ii} and M_{ii} are still consistent with the excitonic picture since the same sub-bands constitute the excitonic states (Fig. 5.2b).

Excitons in carbon nanotubes are characterized by large binding energies, exceeding 700 meV for a 1 nm diameter pristine tube [17]. The excitonic states are therefore stable at room temperature, in contrast to conventional bulk semiconductors where excitons with binding energy of tens of meV can only be observed at low temperatures [18]. The large binding energies result from weak dielectric screening of the Coulomb interaction; the atomically thin structure of the nanotubes offers weak screening to start with, and the 1D geometry further limits any screening effects [19].

These enhanced light-matter interactions give rise to abundant phenomena that can be probed by optical microscopy and spectroscopy, i.e. direct photon absorption [16, 20, 21], elastic/inelastic scattering of photons (Rayleigh/ Raman scattering) [16, 22–24] and luminescence from either optical or electrical excitation [25–31] (fluorescence or electroluminescence). Since optical transitions of SWCNTs are closely related to their geometrical structures, these high-throughput and non-invasive optical techniques have been widely used to characterize their chirality and to probe the physical properties of carbon nanotubes in the past thirty years [13].

Rayleigh scattering, which provides information about the nanotube structure through resonant enhancement of the elastically scattered light when the photon energy matches that of an electronic transition, has emerged as a characterization technique for individual SWCNTs. Rayleigh scattering techniques measure an absolute signal instead of a fractional change and uncover the optical properties of individual nanotubes. Optical absorption, in contrast, is challenging because single tubes change the intensity of the incoming light by $10^{-5} - 10^{-4}$. To solve this, high-contrast polarization microscope has been developed for absorption measurement of individual nanotubes suspended or on substrate. The basic principle and related exploration of absorption/Rayleigh scattering of individual SWCNTs will be discussed in Sect. 5.3.

Photoluminescence excitation (PLE) spectroscopy is a widely used optical technique for studying SWCNTs, especially for performing chirality assignments through determination of the absorption and emission energies [6, 15, 32–35]. In addition, PL spectroscopy is a non-intrusive technique that can be performed on as-grown air-suspended individual nanotubes without further sample processing, ideal for characterizing the pristine material. By automating the measurement system, few thousand nanotubes can easily be characterized, allowing for chirality-on-demand measurements. More details of PLE spectroscopy for air-suspended individual SWCNTs will be reviewed in Sect. 5.4.

Resonant Raman scattering spectroscopy of SWCNTs [24] is regarded as the most powerful characterization tool for SWCNT samples. Even from bulk samples, characteristic peaks of phonon features such as G-band D-band, RBM peaks, and

2D-band (G') can be observed. Because RBM frequency directly corresponds to the diameter of SWCNTs, diameter distribution of bulk samples can be roughly estimated by Raman measurements. However, the strong resonant condition depending on excitation laser sometimes hinders the interpretation. The coupling of excitons and phonons has been the central research topic of physics and also metrology. D-band which comes from the double resonance process possible with defects in the hexagonal lattice is a good indication of the quality of SWCNTs. Spectroscopy of individual SWCNTs makes the chirality dependent G^+, G^- features much clearer. Because transition energy and Raman frequency are sensitive to the surrounding environment, discussion of the environmental effect on Kataura plot is crucially important. Details will be discussed in Sect. 5.5.

Although more than a quarter century's intense research and exploration for their distinctive physical properties and potential applications, spectroscopy of carbon nanotubes remain an active research field with many surprises, and opportunities. Recently, the interlayer coupling, modulation of electronic structure and phononic structure of double-walled carbon nanotubes (DWCNTs) has been explored. In addition, hetero-nanotubes based on SWCNTs, like SWCNT surrounded coaxially by hBN nanotube or transition metal dichalcogenides (TMD) nanotubes, give exciting spectroscopy results. Besides, Luttinger-liquid plasmons has been revealed by near field spectroscopy of both metallic and semiconducting SWCNTs. Furthermore, SWCNTs coupled with silicon photonics devices has been developed and single-photon emission at room temperature has been demonstrated. Such recent advances and prospects of spectroscopy of SWCNT will be discussed detailly in Sect. 5.6.

5.3 Absorption/Rayleigh Scattering

5.3.1 Rayleigh Scattering.

Rayleigh scattering, which relies on elastic scattering of incident light and does not require coupling of incident photons to the phonons in the material, is several orders of magnitude stronger than inelastic Raman scattering [36]. Initially, a tungsten lamp was used as a white light source in Rayleigh spectroscopy for suspended SWCNT bundles, and resonance peaks corresponding to their interband transitions were observed [37]. After that, the use of a laser-based supercontinuum source brought breakthroughs to the field, as it realized parallel collection of spectra over a wide range of wavelengths [16, 22, 23] At the same time, tight focusing of the incident beam and data collection by oblique objectives provided detection of the elastically scattered light over a large solid angle. With the high spatial and spectral resolution, Rayleigh scattering signal from both metallic and semiconducting freely-suspended nanotubes in the visible region can be detected [22, 23]. Sharp resonant peaks arising from transitions between sub-bands were observed in the Rayleigh scattering spectrum (Fig. 5.3b).

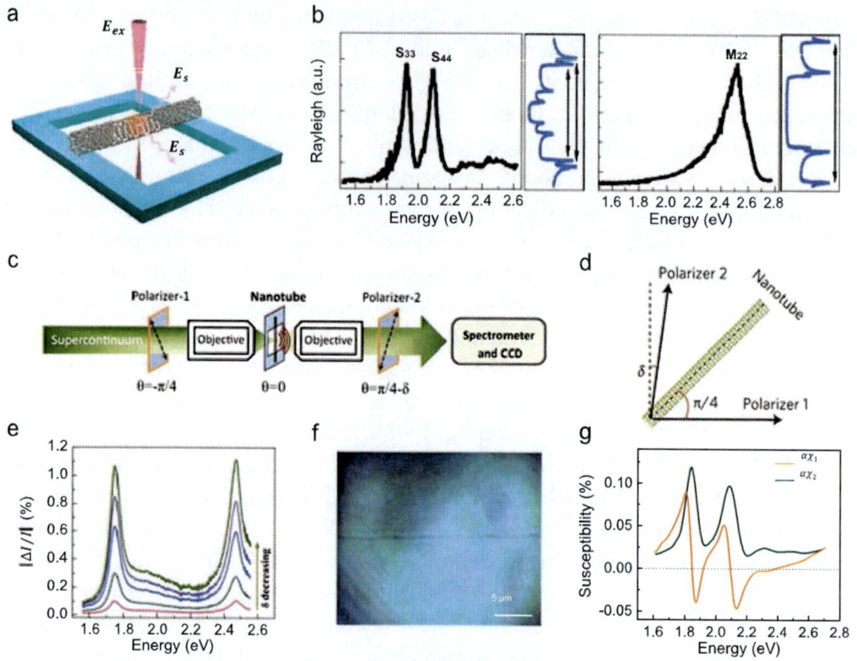

Fig. 5.3 Rayleigh scattering and optical absorption of SWCNTs. a, Scheme of Rayleigh scattering measurement. The supercontinuum incident light (E^{ex}) focuses on suspended individual nanotube and the scattered light (E_s) can be collected. **b,** Rayleigh scattering spectrum of an isolated semiconducting nanotube (left), and a metallic counterpart (right). Scattering is significantly enhanced with the resonance of singularities in the density of states. **c,** Scheme of polarization-optimized homodyne detection for individual nanotube absorption. Two nearly crossed polarizers with a small deviation angle δ were used to control the incident and outgoing light polarization, and the suspended individual nanotubes were placed at the focus of the objectives with an angle of $\pi/4$ with respect to the first polarizer polarization. **d,** Configuration of incident polarizer 1, outgoing polarizer 2 and nanotube for polarization-based optical microscopy. **e,** Homodyne modulation signal ($\Delta I/I$) at various values of deviation angle . With decreasing from 5 to 0.5°, the signal at resonances gradually increases to the 1% level. **f,** Direct optical images of an individual nanotube on silica substrate. **g,** Real ($\alpha\chi_1$, orange line) and imaginary ($\alpha\chi_2$, green line) susceptibility of the SWCNT (19, 11). Figure adapted with permission from: **b,** ref 22, AAAS; **c, e,** ref 20, NAS; **d, f, g,** ref 64, Springer Nature Ltd.

These ground-breaking experiments sparked intense activity on the research of Rayleigh scattering spectra for SWCNTs [38–42]. For instance, measurements show that the polarization dependence of Rayleigh scattering follows the $\cos^2\theta$ form, as expected for dipole emission along the nanotube axis [22]. In addition, direct intertube electronic coupling and dielectric screening bring strong tube-tube interaction, which was directly probed as resonant peak shifts by several tens of meV in the Rayleigh scattering spectroscopy [12]. The backscattering reflection spectrum of suspended individual metallic SWCNTs was collected and an exciton-phonon sideband separated by 0.19 eV from the first optical transition peak was observed, which provided clear evidence of excitons in metallic SWCNTs. Further, by coating an index-matching medium with the substrate, background scattering was minimized and direct true-color Rayleigh imaging of SWCNTs on quartz was realized [40]. Key SWCNT parameters, including the electronic-types and chiral indices, thus can be mapped quickly. Besides, the spatial distribution of the radiation scattered by the nanotubes was revealed and long-range radiative coupling of SWCNTs was demonstrated. Similarly, by coating an interfacial layer of condensed molecules to enhance the local field, Rayleigh scattering of SWCNTs can be greatly enhanced [42]. Benefitting from that, real-time, true-color Rayleigh imaging and spectroscopy on different substrates (SiO_2/Si and quartz) and environment (water and air) have been successfully achieved, offering rapid feedback of key nanotube properties to specific growth conditions [42].

Complementarily, by combining Rayleigh scattering spectroscopy with electron diffraction, the optical transitions and chiral indices can be directly linked with each other [13, 23]. Up to now, a complete optical-transitions-to-chiral-index atlas has been established [13]. With this structure-property relation map, an experimental Kataura plot was further obtained. This atlas also enables the structure characterization by Rayleigh spectroscopy, which facilitates investigation of the electron-phonon coupling, strain-induced effect, controlled growth [43, 44], and mechanical properties of individual nanotubes [45–47].

5.3.2 Absorption Measurement

Optical absorption spectra are critical for various applications of SWCNTs, since optical properties and many optical parameters including absorption coefficient, scattering coefficient, dielectric coefficient, and refractive index, can be derived from them. Given the weak optical absorption of an individual nanotube ($< 10^{-4}$ for a tightly focused laser beam in the visible region), the signal is buried in the background and laser intensity fluctuations. Thus, compared to Rayleigh scattering that measures the response in a dark-field geometry, it is more difficult to determine the optical absorption of an individual nanotube.

Early absorption experiments were carried out on carbon nanotube films, where energy-dependent and polarization-dependent absorption cross-sections were revealed [48–51]. Thereafter, a UV-vis-NIR absorption spectrum of the micelle-separated single-chirality SWCNTs in solution were obtained and used to determine the purity and chirality [9, 52–54]. For individual SWCNTs, detours had to be made to detect the absorption information in the photocurrent excitation spectrum [55–58] or photothermal signal [59], but these techniques suffer from slow laser-frequency scanning and limited spectral range.

The quantitative measurement of the absorption coefficient for an individual CNT was first realized by using the spatial modulation spectroscopy technique [21, 46, 59–61]. By spatially moving the nano-object in and out of the focal spot of a tightly focused light beam, spatial modulation spectroscopy technique induces a modulation of the transmission or reflection intensity. By directly measuring these modulation, the absorption spectrum of an individual nanotube can be obtained. Under each given optical frequency, the detected intensity change ($\Delta I/I$) is the result of the interference between the small optical signal (E_s) of samples and electric field (E_{LO}) of excitation light:

$$\frac{\Delta I}{I} = \frac{|E_{LO} + E_s|^2 - |E_{LO}|^2}{|E_{LO}|^2} = \frac{2|E_s|}{|E_{LO}|}\cos\varphi \tag{5.1}$$

where φ is the relative phase between E_s and E_{LO}. The small $|E_s|^2$ term can be ignored. The resulting mixed modulation signal ($\Delta I/I$), which carries the information (amplitude, phase) of the original desired signal (E_s), can be larger than E_s and then easily detected. Based on this concept, the absorption cross-section of SWCNTs per unit length was successfully determined [20, 21]. Furthermore, by comparing the experimental absorption spectra of the metallic and semiconducting nanotubes, exciton binding energy in metallic SWCNT was deduced to be 50 meV. Besides, by comparing the resonance peaks of the same SWCNT with free-standing samples or those deposited on SiO_2/Si substrate, a large broadening and a red-shift induced by the substrate were revealed.

To enhance the small absorption signal and quantitatively determine the absolute absorption cross-section of individual CNTs in a wide spectral range, the large background signal of excitation light must be suppressed. Utilizing the strong depolarization effect of 1D CNTs, cross-polarization techniques provide the possibility to enlarge the contrast signal by reducing the local excitation electric field (E_{in}) while maintaining the signal (E_{NT}) of SWCNT [13, 41, 48, 62]. As shown in Fig. 5.3c,d, by setting the SWCNT between two perpendicular polarizers (with a small deviation angle δ) at an angle of 45° with the main axes of the polarizers, the excitation light which keeps the incident polarization can be reduced by several orders of magnitude, thus making the weak nanotube signal observable. With a small δ, the greatly enhanced optical contrast $\Delta I/I$ can reach ~1% level (compared with 10^{-4} in conventional transmission methods) and become easily detectable (Fig. 5.3e). Combining with the laser spot profile, the absorption cross-section of many individual CNTs could be quantitatively determined. Furtherly, by extrapolating enough data, an empirical formula was established, which can be used to predict the absorption cross-section spectrum for any given nanotubes [41].

Fundamentally, the strength of the absorption is determined by the imaginary part of the susceptibility, and Rayleigh scattering relies on the square of linear optical susceptibility of the nanotube. Recently, based on the cross-polarization strategy, an elliptical polarization-based optical homodyne detection technique was further demonstrated and the detection of both the real and imaginary parts of $\tilde{\chi}$ of individual carbon nanotubes was realized [63]. By utilizing the interference between incident left-/right-handed polarization beam and nanotube scattering light, two sets of optical spectra containing both χ_1 and χ_2 information with different pre-coefficients can be obtained, which allows one to determine the quantitative value of $\tilde{\chi}$ (Fig. 5.3e). In addition, elliptical polarization light with a relatively large ellipticity and a nearly crossed polarizer were used, which can greatly reduced the background of excitation light, enhancing the optical signal level by about two orders of magnitude and making the extremely weak individual nanotube signal readily detectable.

5.3.3 In-Situ Optical Imaging

Direct visualization of individual carbon nanotubes in ambient conditions is of great significance for their manipulation and applications. Compared with the visualization of SWCNTs with the scanning electron microscope (SEM), transmission electron microscope (TEM), atomic force microscope (AFM), scanning tunneling microscope (STM) and so on, optical imaging of carbon nanotubes with high throughput can be quite more convenient to integrate with other facilities, thus making it highly desirable and long been pursued for decades. Because their nanoscale dimensions are much smaller than the wavelength of visible light, it is generally difficult to observe them by optical microscopes.

One approach to address this challenge is to take advantage of the scattering of other nanoparticles. By depositing Ag, Au, TiO_2 nanoparticles [64, 65], or water droplet onto SWCNTs [66], the visualization and manipulation of individual ultra-long carbon nanotubes under optical microscopes could be realized. However, such methods are generally subject to complicated deposition processes or short imaging times.

The second pathway is by the light-matter interaction process of SWCNTs and high-end scientific cameras. Based on fluorescence, Raman, electro-luminescence and photocurrent process of SWCNTs [67–69], the morphology of individual nanotubes can be successfully mapped by moving the laser focus. However, they are time-consuming. Notably, by using the Rayleigh signal with relatively strong intensity and the strong optical anisotropy, polarized light microscopy was developed to image individual suspended CNTs. In addition, as mentioned before, true-color imaging of SWCNTs on a substrate, which uses refractive index matching to reduce the reflected light of the substrate has also been reported. Furthermore, polarization-based optical microscope in reflection configuration has been successfully constructed and high-throughput optical imaging and in-situ spectroscopy of individual CNTs in devices have been realized (Fig. 5.3f). By manipulating the polarization, the reflection field was strongly reduced, meanwhile, the nanotube field was kept so that the optical contrast could be sufficiently enhanced. As such, optical contrast spectra of a single CNT can be acquired in situ and its chirality can be determined accurately with the aid of an experimental transition-chirality atlas. So far, this technique has been applied to give feedback to the growth, monitoring device performances, and observation of carbon nanotube etching process [44, 70, 71].

5.4 PL Microscopy of Individual-Suspended Carbon Nanotubes

5.4.1 Air-Suspended Carbon Nanotubes

Air-suspended SWCNTs grown over trenches on bare Si substrates are used for PL microscopy measurements [72]. The trenches, as well as sample alignment marks, are formed by electron beam lithography and dry etching, where the widths of the trenches range from 0.2 to 6.8 μm. Catalyst areas are patterned by another electron beam lithography step as shown in Fig. 5.4a. As a standard catalyst for nanotube growth, 20 mg of Fe(III) acetylacetonate and 100 mg of fumed silica dispersed in 40 g of ethanol are used. The catalyst particles are deposited using spin-coating and lift-off techniques, and the samples are heated in air at 400 °C for 5 min.

SWCNTs are synthesized by alcohol chemical vapor deposition [73]. Samples are brought into a quartz tube with an inner diameter of 26 mm, and its internal pressure is kept at 110 kPa with a gas mixture of 97% Ar and 3% H_2 flowing at 300 sccm. The temperature is ramped to 800 °C over 15 min by an electric furnace, and the gas flow and the temperature are kept constant for another 15 min to reduce the catalyst metal. Finally, ethanol vapor is delivered into the quartz tube for 10 min by switching the gas flow to bubble through a bottle of ethanol [74]. Figure 5.4b is a scanning electron micrograph of a sample after the nanotube growth process.

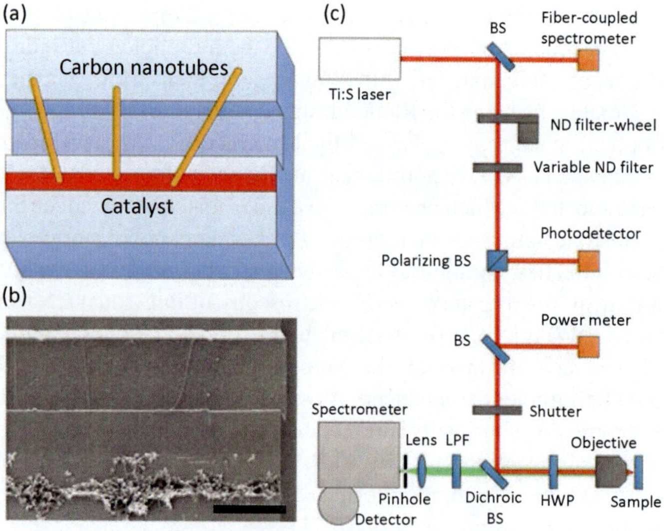

Fig. 5.4 **a** A schematic of a sample. 10 mm × 10 mm chips with 144 trenches with a length of 900 μm are used. **b** A scanning electron micrograph of a typical sample. The scale bar is 2 μm. **c** A schematic of the optical setup. The thin line represents the excitation beam and the thick line indicates the PL collection path. Original data presented in [72]

5.4.2 PL Spectroscopy of Individual Nanotubes

An automated confocal PL measurement system (Fig. 5.4c) has been constructed for characterization of a large number of nanotubes [72]. In the set-up, a continuous-wave Ti:sapphire (Ti:S) laser is used for excitation, whose wavelength is controlled by a motorized linear actuator attached to the birefringent filter of the laser. The output beam passes through a polarizing beam splitter (BS) to clean the polarization and allow the collection of the reflected beams from the sample. The beam is split into two paths, and one of the beams enters a power meter. Calibration has been performed so that the actual excitation power on the samples can be obtained. With a feedback control using the variable neutral-density (ND) filter and the power meter, the excitation power can be tuned within an error of less than 1%. Passing through a shutter used for background subtraction, the transmitted beam is directed towards the sample by a dichroic BS, and then its polarization is rotated by a half-wave plate (HWP) mounted on a motorized rotation stage. An objective lens is used to focus the excitation beam and to collect the emission signals from the nanotubes. The wavelength-dependent $1/e^2$diameter of the focused laser has been characterized by performing PL line scans perpendicular to a suspended nanotube.

The samples are mounted on an automated three-dimensional stage with a travel range of ± 10 mm and a resolution of 50 nm, which is used for focusing as well as sample scanning. PL emitted from the nanotubes transmits the dichroic BS, and a long pass filter (LPF) with a cut-on wavelength at 950 nm is used for further laser rejection. PL from the sample is focused by a lens with a focal length of 50 mm, and it passes through a confocal pinhole with a diameter of 150 μm, corresponding to an aperture with 5.4 μm diameter at the sample image plane. Liquid-nitrogen-cooled InGaAs photodiode array attached to a 300-mm spectrometer with a 150 lines/mm grating blazed at 1.25 μm is used to obtain the PL spectra, which provides a spectral resolution of ~1 meV.

The reflected beam from the sample traces back the same path as the excitation beam and is detected by a photodiode. By performing reflectivity scans, position offsets and rotation angles of the samples can be determined from the alignment marks. In addition, by bringing the laser focus at three different positions on the surface, tilt angles are obtained. Coordinate transformation can be performed from the results of these measurements, allowing the entire area of the samples to be scanned while keeping it in focus. All measurements are performed at room temperature in air unless otherwise noted.

5.4.3 Statistical Characterization of Air-Suspended Nanotubes

The suspended nanotubes are located by line scans along the trenches [Fig. 5.5a], and PLE measurements are performed for each nanotube. The PLE maps of individual

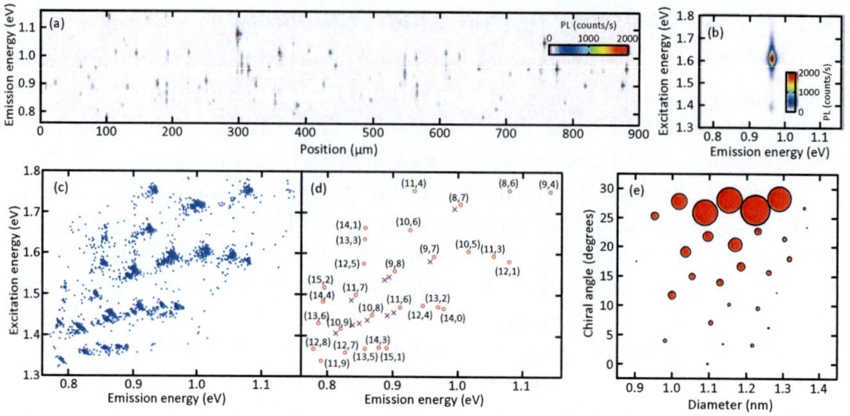

Fig. 5.5 **a** A typical result of a trench scan with $P = 50$ μW and $E_{ex} = 1.59$ eV. **b** A typical PLE map of a (9,7) nanotube with $P = 1.5$ μW. **c** PLE peak positions of 3736 individual nanotubes. **d** Averaged peak positions for each chirality. Open circles represent main spots, and cross marks indicate satellite spots. **e** Chirality distribution as a function of tube diameter and chiral angle. The area of the circles represents the population. Original data presented in [72]

nanotubes present distinct peaks in both emission and excitation energies as shown in Fig. 5.5b, where S_{11} and S_{22} energies of the nanotubes are obtained from Lorentzian fits. As PLE maps with multiple peaks and significant broadening may come from bundled tubes [75] or defect states, such nanotubes are excluded from further measurements. These scans are performed automatically overnight, where typically 36 trench scans are done within 9 h while 300 PLE maps are taken within 10 h. The positions and chiralities for thousands of nanotubes are recorded into a list, allowing for a statistical analysis [72]. Such a list is also utilized for chirality-on-demand measurements later (Sect. 5.6.4).

In Fig. 5.5c, peak positions in the PLE maps are plotted for all of the measured individual nanotubes, where high density spots corresponding to different chiralities can be seen. Chiralities are assigned by utilizing the results on ensembles of nanotubes [32–34], and the averaged peak positions for each chirality are summarized in Fig. 5.5d and Table 5.1. Interestingly, the PLE peak distribution in Fig. 5.5c shows satellite spots with slightly lower energies. The energy shifts between the main spots and the first satellites do not depend much on chirality, and the average values are 8.1 and 12.3 meV for S_{11} and S_{22}, respectively. For chiralities with large population such as (9, 8) and (10, 8), there are second and third satellite spots similar to the first satellite spots but with decreasing populations. As no apparent differences in PLE maps between the main spot and the satellites are observed except for the slight redshifts, these satellites are assigned to bundles of nanotubes with the same chirality. This interpretation is consistent with the redshifted values reported in [34], where samples without catalyst patterning are used and more bundles are expected. It is also interesting that the distributions of the peaks for the same chiralities are not isotropic, but show up as elongated ellipses corresponding to anticorrelated S_{11} and

Table 5.1 Average S_{11} and S_{22} peak energies for individual air-suspended CNTs obtained from Lorentzian fits of their PLE maps. The error values are standard deviations. Original data presented in [72]

	S_{11}		S_{22}	
(n, m)	(nm)	(meV)	(nm)	(meV)
(8,6)	1148.0±4.4	1080.0±4.2	707.0±2.4	1753.6±6.0
(8,7)	1235.1±5.0	1003.8±4.0	720.9±2.0	1719.9±4.7
(9,4)	1085.3±3.5	1142.3±3.7	708.0±5.6	1751.0±13.9
(9,7)	1286.8±3.9	963.5±2.9	779.0±2.6	1591.6±5.3
(9,8)	1372.4±3.8	903.4±2.5	796.8±2.6	1556.1±5.1
(10,5)	1219.8±4.1	1016.4±3.4	773.0±3.3	1604.0±6.9
(10,6)	1337.7±3.7	926.8±2.6	748.2±2.1	1657.1±4.8
(10,8)	1427.8±4.3	868.3±2.6	855.7±2.5	1448.8±4.2
(10,9)	1511.1±5.1	820.4±2.8	875.4±2.2	1416.3±3.5
(11,3)	1174.5±3.5	1055.6±3.1	778.1±3.0	1593.4±6.1
(11,4)	1328.2±4.7	933.4±3.3	707.5±2.3	1752.4±5.7
(11,6)	1360.2±3.9	911.5±2.6	844.7±2.6	1467.8±4.5
(11,7)	1470.1±3.0	843.3±1.7	827.6±2.0	1498.1±3.7
(11,9)	1568.0±5.5	790.7±2.8	927.3±2.8	1337.0±4.1
(12,1)	1148.6±2.9	1079.4±2.7	784.8±2.5	1579.8±5.1
(12,4)	1310.0±3.0	946.4±2.2	842.1±2.1	1472.2±3.7
(12,5)	1448.7±3.6	855.8±2.2	787.7±1.8	1573.9±3.7
(12,7)	1498.7±4.3	827.3±2.4	913.4±1.9	1357.3±2.8
(12,8)	1591.2±3.1	779.2±1.5	907.6±2.2	1366.0±3.4
(13,2)	1278.8±2.4	969.5±1.8	843.9±2.1	1469.1±3.7
(13,3)	1446.6±4.5	857.0±2.6	758.7±3.1	1634.1±6.6
(13,5)	1446.1±3.5	857.4±2.1	906.7±1.6	1367.4±2.4
(13,6)	1576.1±3.8	786.6±1.9	867.7±1.8	1428.8±3.0
(14,0)	1267.3±2.5	978.3±1.9	846.1±1.6	1465.3±2.8
(14,1)	1445.5±3.8	857.7±2.2	745.9±2.2	1662.1±4.9
(14,3)	1411.1±3.4	878.6±2.1	905.0±1.7	1369.9±2.6
(14,4)	1561.9±3.6	793.8±1.8	837.1±1.7	1481.1±3.1
(15,1)	1392.1±2.3	890.6±1.5	905.7±2.1	1369.0±3.1
(15,2)	1558.6±1.8	795.5±0.9	817.7±3.1	1516.2±5.8

S_{22}. This cannot be explained by dielectric screening effects, but seems consistent with the effects of slight strain [76–78] or bending [79].

The automated scans allow us to determine the chirality distribution of semiconducting nanotubes in the samples. Trench scans are repeated three times using laser photon energies E_{ex}= 1.46, 1.59, and 1.75 eV to locate nanotubes with different S_{22} resonances. The results of the scans are carefully compared to verify that the same nanotube is only counted once and chiralities are assigned by performing PLE measurements. The chirality distribution is obtained by counting the number of nanotubes, and the results are plotted in Fig. 5.5e as a function of nanotube diameter and chiral angle. It is clear that the nanotubes with larger chiral angles show larger population, in agreement with reports of chirality-dependent growth rates [61, 80–84]. This result is not dependent on physical parameters such as PL quantum yield [81, 82] or Raman scattering cross section [84], although PL measurements can only detect semiconducting nanotubes and the spectral range is limited by the capabilities of our laser and detector.

5.5 Raman Scattering

5.5.1 G-band and Resonance Features

The Raman scattering and PL can be measured simultaneously from an air-suspended SWCNT after the assignment of chirality by PLE mapping as described in the previous section. An example of Raman scattering from (9, 7) SWCNT is shown in Fig. 5.6 (top spectrum). The SWCNT was grown on Si pillars. The PLE mapping similar to the previous section was performed in order to identify the chirality [85] as in Fig. 5.6b. The clear single RBM and G mode features are obtained. Since these spectra are measured under the water vapor pressure of 2.67×10^3 Pa, RBM frequency is blue shifted about $7\,cm^{-1}$, and PL emission wavelength is red shifted about 35 nm [85]. The Raman spectrum of (9, 7) SWCNT is compared with isolated SWCNT on SiO_2 substrate which were transferred from horizontally aligned SWCNTs on a crystal quartz substrate [86]. Because of the aligned growth, most of SWCNTs are free from bundling or crossing. Except for the environmental effects of SiO_2 and adsorbed water, pristine features of individual SWCNTs can be expected. The bottom Raman spectrum in Fig. 5.6a is from random and bulk SWCNTs grown on catalysts supported by zeolite particles [87]. In addition to the broad range of chirality distribution, certain broadening of each peak causes the complexity of interpretation of Raman peaks.

The G-band originates from an in-plane stretching mode of carbon-carbon bonds in graphitic materials [88]. For graphite, since the G-band comes from phonons with E_{2g} symmetry, which are equivalent longitudinal and transverse optical phonons at Γ point, the G-band appears as only one peak around $1580\,cm^{-1}$. In the contrast to the graphite or graphene, the G-band from SWCNTs appear around $1590\,cm^{-1}$as multiple peaks [89]. The G-band of SWCNTs composed of multiple peaks was explained to be the superposition of peaks from phonons with different symmetry, such as $2A$, $2E_1$ and $2E_2$ modes [89]. Even more complex G-band is often discussed from bundled SWCNT samples. Especially, two most prominent peaks among the G-band are named $G^+$and G^-peaks. Here, lower and higher frequency peaks are defined as $G^-$and G^+, respectively. The discussion was not straightforward for bulk sample as in the bottom spectrum in Fig. 5.6a. Raman scattering spectra from individual semiconductor SWCNT in Fig. 5.6 are quite simple. There are only symmetric G^+ and

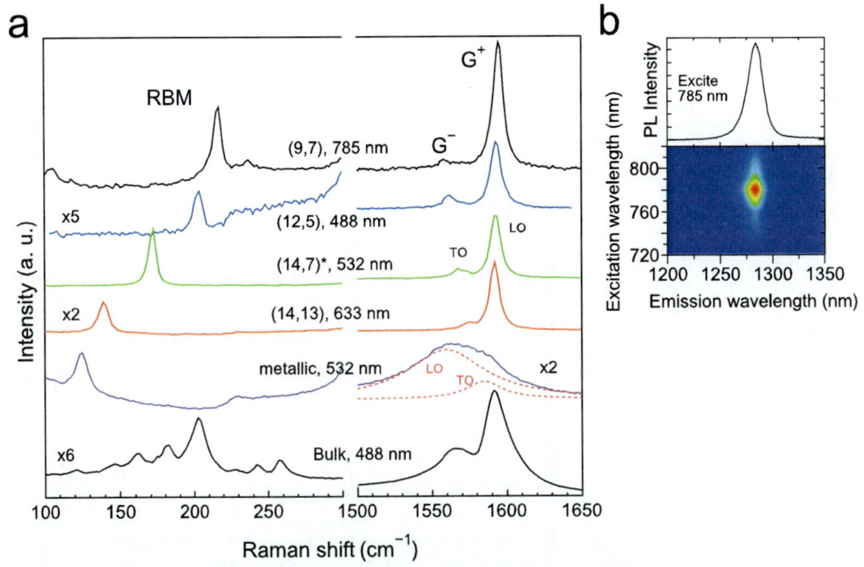

Fig. 5.6 **a** Raman scatting from a suspended (9, 7) SWCNT compared with various individual SWCNT on SiO_2 substrates and bulk SWCNTs. **b** PL of (9,7) SWCNT simultaneously measured with Raman scattering. The (9, 7) SWCNT is kept in water vapor at 2.67×10^3 Pa. The excitation laser was 785 nm. The PLE map in (**b**) is measured for the identification of chirality as in [85]. Original data are courtesy of Prof. Shohei Chiashi at UTokyo. Original Raman spectra for individual SWCNT on SiO_2 substrate transferred from horizontally aligned samples on crystal quartz are from [86]. Chirality of SWCNTs on SiO_2 substrate are determined by Kataura plot and systematic changes in the ratio of intensities of RBM peaks and G-band as in [86]. The assignments of (14, 7)* is hence tentative. Raman spectrum for metallic SWCNT was averaged along a SWCNT. Here, dotted lines LO and TO are obtained by the decomposition of G-band by 2 Lorentzian functions. The bulk SWCNTs is grown by alcohol catalytic CVD technique in [87]

small G^-peaks. The features are common to dispersed and chirality sorted nanotubes in solution [90] and previous suspended SWCNTs in air [91].

Raman spectra of chirality sorted semiconductor SWCNT are shown in Fig. 5.7 from [92]. Just like air-suspended SWCNT as in Fig. 5.6, G-band of single chirality semiconductor SWCNT exhibits only $G^+$and G^-peaks [93]. Because the theory states that the intensity of A mode peaks are strong [94], the G^+ and $G^-$are usually assigned to A symmetry phonons. The feature of $G^+$and G^-peaks from individual or chirality sorted SWCNTs in solution are investigated in detail and it is found that the intensity ratio between RBM, $G^+$and G^- peaks depend on both d_{tube} and the chiral angle [92].

In the case of semiconducting SWCNTs, the $G^+$and G^-peaks, which originate from LO and TO phonons at the Γ point respectively, are quite sharp. The Raman shift of the G^+peak (ω_{G+}) (LO) does not depend on the tube diameter (d_t) because the curvature effect is relatively small for longitudinal phonons. On the other hand, that of the G^- peak (ω_{G-})(TO) clearly depends on the tube diameter due to the curvature

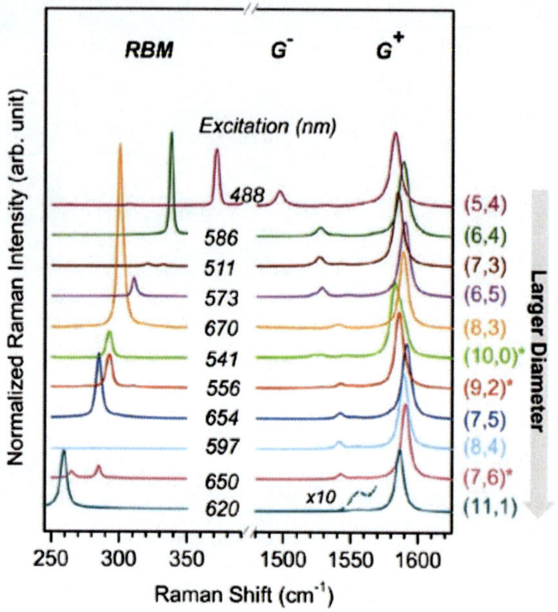

Fig. 5.7 G^+, G^- and RBM peaks of chirality sorted semiconducting SWCNTs. reproduced from [92] with permission of Copyright 2016, American Chemical Society

effect [95]. Therefore, the difference of their Raman shift ($\omega_{G+} - \omega_{G-}$) is roughly expressed by the tube diameter ($\omega_{G+} - \omega_{G-} = C_S/d_t^2$, C_S= 47.7cm^{-1}nm^2) [96]. In detail, ω_{G-} depends on both the diameter and chiral angle [97], as shown in Fig. 5.7.

For metallic SWCNTs, sharp $G^+$and broad G^-peaks are observed and they correspond to TO and LO phonons at the Γ point, respectively. Note that earlier references described the $G^+$and G^- peaks correspond to LO and TO phonons even for metallic SWCNTs. The LO phonon of metallic SWCNTs is softened due to Kohn anomaly [98], and the redshifted and broad (asymmetric) G^-peak of metallic SWCNTs is also called Breit-Wigner-Fano (BWF) peak [99]. Because of the Kohn anomaly, the G-band of metallic SWCNTs is drastically modulated by changing the Fermi energy [98, 100–102].

The density of states of SWCNTs exhibits almost discrete energy levels due to van Hove singularity and their optical transition energy (E_{ii}) corresponds to the energy gap between two energy states. Raman scattering spectra from SWCNTs can be strongly enhanced with resonance excitation. If E_{ii} is equal to the energy of incident light (E_{ex}, incident resonance) or that of the scattered light ($E_{ex} \pm E_{phonon}$, scattered resonance), Raman scattering intensity will be significantly boosted. Here, a Raman process with E_{ex}+ E_{phonon} and E_{ex}- E_{phonon} corresponds to anti-Stokes and Stokes scattering, respectively. Resonance Raman effect [103] is expressed by

$$I(E_{ex}) \propto \left| \frac{1}{(E_{ex} - E_{ii} + i\Gamma)\left(E_{ex} \pm E_{phonon} - E_{ii} + i\Gamma\right)} \right|^2 \tag{5.2}$$

where Γ is a resonance window. Γ is a damping constant and is related to the finite lifetime of the intermediate state. In Raman excitation profiles of the G-band, two resonance peaks, which correspond to the incident and scattered resonances, clearly appear with $\Gamma = 26 - 43$ meV [104] for chirality separated SWCNT.

5.5.2 RBM Peaks

RBM peaks appear in the lower wavenumber region. Raman shift of RBM peaks (ω_{RBM}) is inversely proportional to the SWCNT diameter d_{t} [105, 106]as follows,

$$\omega_{\mathrm{RBM}} = A/d_{\mathrm{t}}. \tag{5.3}$$

Therefore, RBM peaks are often used to analyze the diameter distribution of SWCNT samples. However, ω_{RBM} is not exactly expressed by (3), because of the environmental effects on ω_{RBM}. Considering the environmental effect, the following equation,

$$\omega_{\mathrm{RBM}} = A/d_{\mathrm{t}} + B \tag{5.4}$$

is often used. This simple equation is useful for analysis of the SWCNT diameter, although the values of A and B vary depending on the environment. A discussion of more sophisticated dependence incorporating the environmental effect can be found in [107].

RBM peaks are strongly enhanced by resonance Raman effect and the resonance Raman effect is also expressed by Eq. (5.2). Spectra of RBM peaks of bulk sample change with the excitation wavelength. Each peak comes from different chirality and the intensity also clearly depends on the chirality. In general, SWCNTs with smaller chiral angle (near zigzag) exhibit stronger intensity for the RBM peaks [108]. Additionally, the relationship between the intensity of $G^+$and RBM peaks is investigated for chirality separated SWCNT samples and it shows chirality dependence [92].

5.5.3 Kataura Plot

As mentioned above, the intensity of RBM peaks is strongly enhanced by resonance Raman scattering effect. The phonon energy of RBM peaks is approximately a few tens of meV. When E_{ex} is in the visible or near infrared range, E_{phonon} of RBM peaks is quite small compared with E_{ex}. Therefore, it is usually impossible to distinguish the incident and scattered resonances in Eq. (5.2) and only E_{ii} is simply regarded as the resonance energy in RBM peaks. The relationship between ω_{RBM} and E_{ii} is called the Kataura plot [4]. Figure 5.8 shows an example of Kataura plot, which is a compilation of experimental data [13, 84, 109–113]. Because both ω_{RBM} [105, 106] and E_{ii} [114] are roughly proportional to $1/d_{\mathrm{t}}$, the plots in Fig. 5.8 appear along lines from the lower

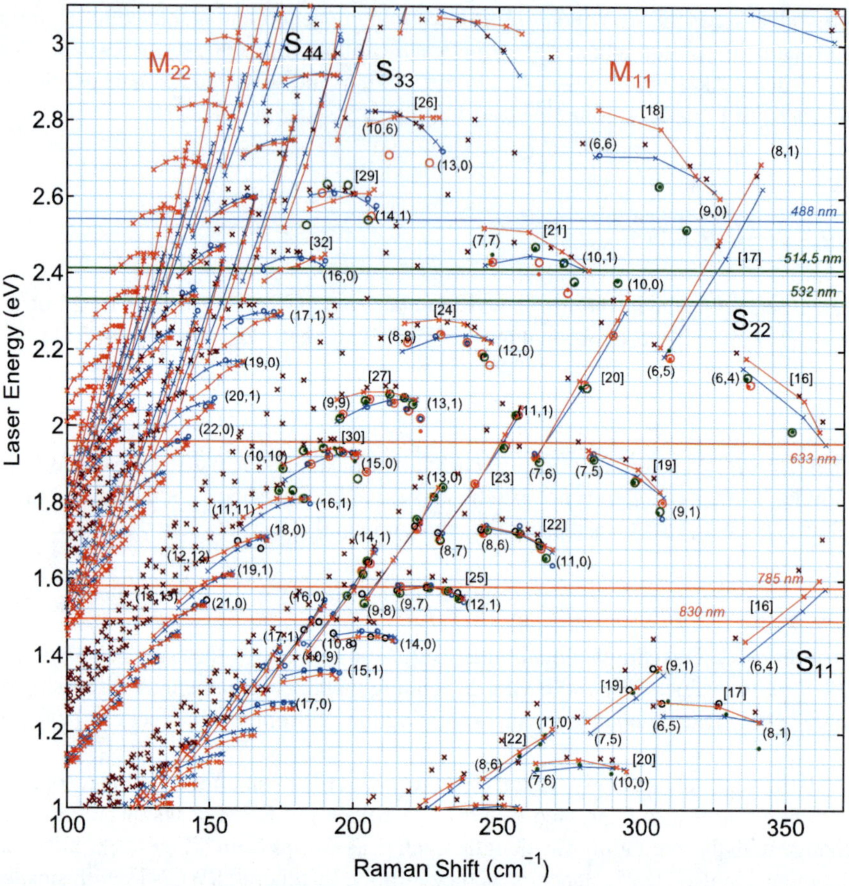

Fig. 5.8 Empirical Kataura plots of SWCNTs. Blue open circles and blue cross marks connected with blue lines (S_{11}, S_{22}, M_{11}, S_{33}, S_{44}) are measured and empirical formula, respectively, from Araujo et al. [110]. Red cross marks connected with red lines (S_{11}, S_{22}, M_{11}, S_{33}, S_{44}, M_{22}) are from nanotube atlas by Liu et al. [13] with (A, B) in Eq. (5.4) set as (217.8, 15.7). Brown cross marks are calculated from Araujo et al. [113]. Green open circles from Telg et al. [111], red open circles from Fantini et al. [109] and red solid circle from Jorio et al. [84] are measured from dispersed SWCNTs in solution. Green solid circles are measured from suspended SWCNTs from Michel et al. [112]. Chiral index (n, m) and family pattern branches $[2n{+}m]$ are indicated

left to the top right. Here, the S_{ii} and M_{ii} are the i-th optical transition energy of semiconducting and metallic SWCNTs, respectively. In Fig. 5.8, 6 different optical transition energies are shown, such as S_{44}. Because E_{ii} depends on both d_{t} and the chiral angle (θ), the plots do not appear just along the lines and are slightly spread. In Fig. 5.8, the plots with the same value of $2n{+}m$ are connected with line segments and they show the so-called "family pattern" [115]. In Fig. 5.8, some of plots are denoted with chiral index (n,m) and some of branches of family pattern are denoted with $[2n{+}m]$. As mentioned in Sect. 5.2, semiconducting SWCNTs with $(2n + m)$ mod 3 =

1 and 2 are called 'type-I' ('SI' or 'mod 2') and 'type-II' ('SII' or 'mod 1') SWCNTs, respectively. As an example, (7, 5) SWCNT schematically shown in Figs. 5.1 and 5.2a is type I ($2n$+m=19). As shown in Fig. 5.2a, a cutting line corresponding to S_{11}, S_{22} for type I SWCNT locate in Γ and M direction, respectively. The direction is opposite for type II. Because the slope of valence-conduction energy gap near K point is sharper in Γ direction than M direction, S_{11} branches spread to larger energy side for type I and smaller energy side for type II. For S_{22}, the spread direction is opposite. Some other properties of SWCNTs show type-dependence. For example, E_{ii} depends on strain and whether E_{ii} increases or decreases is determined by the type [116].

As shown in Fig. 5.8, the empirical Kataura plot proposed by Araujo et al. (2007) [110] match reasonably well with experimental resonant Raman data obtained from micelle-suspended samples by Telg et al. [111], Fantini et al. [109], and Jorio et al. [84]. The air-suspended SWCNTs data from Michel et al. [112] also reasonably agree with the empirical Kataura plot. This is rather surprising because the original SWCNT sample used by Araujo et al. [110] was vertically aligned SWCNT forest [49]. Later, we understand that randomly-oriented, less-bundled, and small diameter SWCNTs suspended in the forest are responsible to resonant Raman signals. The empirical equation in atlas of carbon nanotubes in Liu et al. [13] is also compared in Fig. 5.8. Here, the Raman shift was estimated with the same relation as Araujo et al. [110]. Both empirical relations agree well with slight deviations for S_{33} and S_{44}. Resonant Raman measurements of relatively large diameter and chirality assigned SWCNT are essential to verify which relation is more suitable for better Kataura plot for S_{33} and S_{44}. As in Chiashi et al. [85], by fully removing water absorption on SWCNTs, a considerable red-shift of Raman frequency is observed. The empirical Kataura plot proposed by Araujo et al. [113] using the Eq. (3) is known to fit better for vacuum condition [107].

Based on the accurate Kataura plot, it is possible to assign the chirality from measured RBM peaks. When E_{ii} is close to the E_{ex}, the RBM peaks are observed. The resonance window (Γ) depend on SWCNT samples, such as SDS wrapped SWCNTs in solution (Γ = 60meV) and bundled SWCNTs (Γ = 120 meV) [109]. According to Kataura plots, RBM peaks of bulk sample can be roughly assigned. For example, RBM peak around 200 cm^{-1} and around 250 cm^{-1} in Fig. 5.6a (bottom spectrum) corresponds to semiconducting SWCNTs with $2n + m = 29$ and metallic SWCNTs with $2n + m = 21$, respectively.

5.6 Recent Advances

5.6.1 Inter-Tube Coupling in Double-Walled Carbon Nanotubes

DWCNTs consist of two coaxially aligned SWCNTs, in which the wall can be either semiconducting or metallic depending on its chiral index. However, properties of

DWCNTs are not just a superposition of those of the constituent SWCNTs but can be strongly modulated by (i) interlayer van der Waals mechanical and electronic interactions, mainly depending on the wall-to-wall distance [117–120] and (ii) Moiré interference, mainly depending on the relative chiral angles [121–123]. Early theory predicted that strong wavefunction mixing within two commensurate semiconducting walls could even induce a semiconductor-to-metal transition [124]. However, commensurate DWCNTs have never been observed experimentally because it is almost impossible to have two commensurate SWCNTs with the radius difference matching the tube-tube separation in a DWCNT.

Especially, the interlayer interaction of incommensurate DWCNTs is indicated by the emergence of new and/or collective shifted optical modes in the Raman spectrum [117–120] the modification in the emission PL [125] and significant shifts (−50 to 200 meV) of optical transitions in the absorption spectrum [117]. These experimental findings are also supported by theoretical results that electron wavefunctions between incommensurate inner and outer-wall nanotubes can also hybridize and even weak van der Waals interlayer coupling in incommensurate DWCNTs can slightly change the band structure of pristine SWCNTs [121].

In contrast to the weak-coupling cases, a strong-coupling effect in a structure-identified 1D Moiré superlattice is also experimentally observed and identified [122]. A drastic change of the optical transition in Rayleigh scattering spectrum is found in a DWCNT with chirality (12,11) @ (17,16). The marked change originated from the strong inter-tube coupling effect in the Moiré superlattice formed by two nearly armchair nanotubes, which remarkably changes the electronic band structure and selection rules of optical absorption from the simple sum of the constituent pristine nanotubes [123]. This observation provides valuable insights on inter-tube coupling and may lead to an extended exploration of rich Moiré physics in 1D structures such as strongly correlated physics, Moiré excitons, and superconductivity.

5.6.2 Hetero-Nanotubes

In addition to DWCNTs, SWCNTs can be coaxially surrounded by hBN nanotubes and transition metal dichalcogenides (TMD) nanotubes. Those hetero-nanotubes based on SWCNTs will provide excitingly rich spectroscopy data. A new coaxial nanotube structure has been synthesized, in which a mono- or few-layer hexagonal boron nitride nanotube (BNNT) seamlessly wraps around a SWCNT; SWCNT@BNNT [126]. These nanotubes are named as 1D van der Waals hetero-nanotubes, because no correlation between the chiral angle of inner SWCNT and outer BNNT for 'double-walled' SWCNT@BNNT were found. We have further developed the 1D coating CVD of transition metal dichalcogenide nanotubes (TMD-NT), such as MoS_2 nanotubes [126]. As shown in Fig. 5.9, these nanotubes can be labeled as SWCNT@BNNT@MoS_2. SWCNT from SWCNT@BNNT can be removed by a gentle oxidation process because BNNTs are thermally more stable than SWCNTs. Hence, the inner diameter of produced BNNTs could be determined by the original

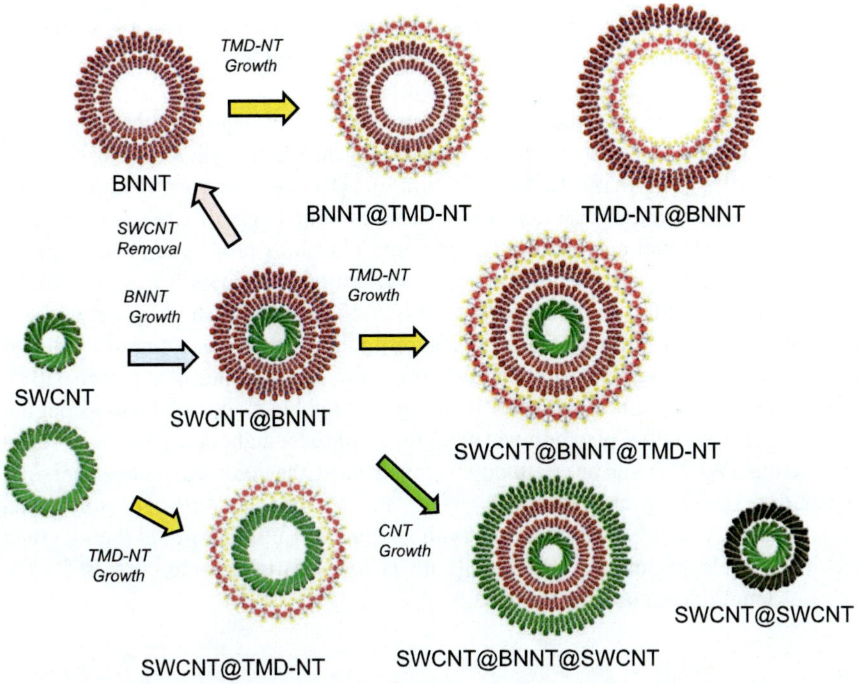

Fig. 5.9 Various structures of hetero-nanotubes

SWCNTs. Subsequent MoS_2 CVD leads to BNNT@MoS_2. The hetero-nanotubes are further characterized by HR-TEM, STEM-EELS, absorption, Raman, and PL spectroscopy (a bright PL for BNNT@MoS_2). From DFT calculations, the 2D MoS_2K point direct band gap is expected in MoS_2 nanotubes with a larger diameter such as 5 nm. For the hetero-nanotube SWCNT@BNNT@MoS_2, the PL from MoS_2 was quenched by inner SWCNT (metallic or small band gap semiconductor). The electronic and phononic structures are expected to be modulated by inter-tube coupling. The inter-tube exciton can also be expected by choosing appropriate semiconductor SWCNT and TMD-NT.

5.6.3 Luttinger-Liquid Plasmon

Quantum-confined electrons in SWCNTs behave as a Luttinger liquid, a strongly correlated electronic matter distinctly different from the quasi-free electrons described by the Fermi liquid [127, 128]. Many electrical transport and photoemission measurements have shown the presence of Luttinger liquid in SWCNTs [129]. Due to this strong quantum confinement, Luttinger-liquid plasmons in SWCNTs with a diameter of 1 nm should persist to visible frequencies before the first intersubband transition appearing. Besides, the forbidden backscattering of Dirac electrons, evidenced by ballistic transport up to micrometer lengths in metallic SWCNTs, can lead to strongly confined but low-loss Luttinger-liquid plasmons.

Recently, the first reported optical observation of Luttinger-liquid plasmons in SWCNTs using infrared scattering-type scanning nearfield optical microscopy (s-SNOM) [130] shows that a Luttinger liquid of 1D Dirac electrons in carbon nanotubes exhibits quantum plasmons which behave qualitatively differently from classical plasmon excitations. The Luttinger-liquid plasmons propagate at 'quantized' velocities that are independent of carrier concentration or excitation wavelength and simultaneously exhibit extraordinary spatial confinement and high quality factor. Following the observation of linear Luttinger-liquid in carbon nanotubes, nonlinear Luttinger-liquid of semiconducting SWCNTs was experimentally explored [131]. With carrier density controlled by electrostatic gating, in situ infrared nano-imaging results have shown that plasmon wavelength, amplitude, and quality factor in semiconducting SWCNTs can be continuously tuned due to the nonlinear band dispersion. In addition, both the propagation velocity and the dynamic damping of plasmons can be fine-tuned, which is well captured by the nonlinear Luttinger liquid theory. Such Luttinger-liquid plasmons could enable novel low-loss plasmonic circuits for the subwavelength manipulation of light.

5.6.4 Coupling to Photonic Structures

Advances in silicon photonics have been expanding the capabilities of monolithic photonic circuits, and the integration of nanoscale emitters would allow for further scaling and increased functionality. In this regard, CNTs are promising because they can be directly synthesized on silicon [132] and they are telecom-band emitters [25, 32] that do not suffer from absorption loss by silicon. Furthermore, nanotube emitters operate at room temperature and they can be electrically driven [27, 30, 133–135].

To utilize such characteristics in monolithic optical circuits, coupling to photonic structures is essential. Microcavities, in particular, offer quantum electrodynamical effects for increased performance and functionality by confining electric fields into a small mode volume. Exciton polaritons can form when nanotube emission is strongly coupled to a cavity [136], whereas the interaction of CNTs with cavities results in an increased spontaneous emission rate in the weak-coupling regime [11, 137–140].

Here PL from individual CNTs are coupled to silicon photonic crystal nanobeam cavities with ultralow mode volumes [11]. The excellent optical properties of as-grown air-suspended CNTs [30, 141–143] are utilized, and they are integrated with specially designed cavities with large fields in the air, distinctly different from the standard dielectric-mode cavities [144–148]. The fabricated devices are shown as a schematic in Fig. 5.10a. Catalyst particles are placed across a trench from the cavity, and chemical vapor deposition is performed to grow CNTs onto the cavities [138]. An electron micrograph of a device after nanotube growth is shown in Fig. 5.10b.

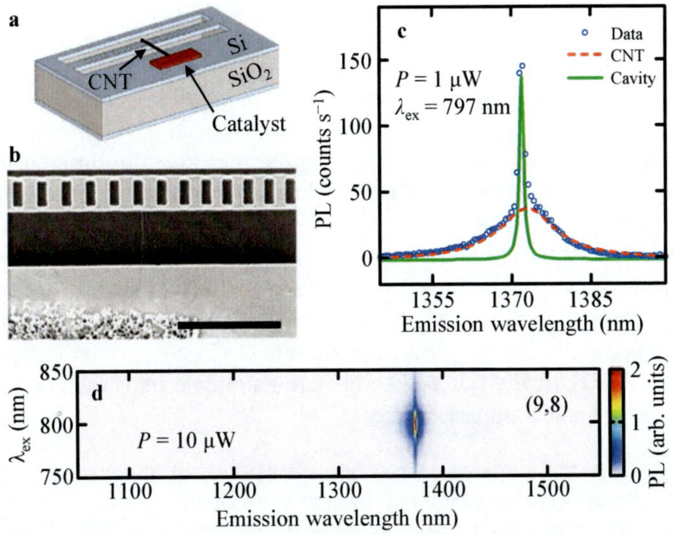

Fig. 5.10 An individual CNT coupled to a nanobeam cavity. **a** A schematic of a device. **b** Scanning electron microscope image of a device with a suspended nanotube. Scale bar is 2 μm. **c** Typical PL spectrum of an air-mode device coupled to a nanotube. The dots are data and the lines are Lorentzian fits. **d** PLE map of the device shown in (**c**) taken with $P = 10$ μW and the laser polarization perpendicular to the nanobeam. Original data presented in [11]

In Fig. 5.10c, a PL spectrum from one of such devices taken with an excitation power $P = 1$ μW and an excitation wavelength $\lambda_{ex} = 797$ nm is shown. On top of the broad direct emission from the nanotube, there is a very sharp peak which is the cavity mode, indicating that the nanotube emission is optically coupled to the cavity. To further characterize the device, PLE spectroscopy is performed (Fig. 5.10d). A single peak is observed in the PLE map, demonstrating an isolated single nanotube. Using tabulated data [34], the chirality is determined to be (9,8). The intensity of the sharp cavity mode is maximized at the same wavelength as the E_{22} resonance of this tube, showing that the absorption originates from the nanotube and that the cavity mode is excited by the nanotube emission.

These results demonstrate the feasibility of integrating telecom-wavelength nanotube emitters in silicon photonics, opening up further possibilities for scaling down monolithic photonic circuits. By interfacing the cavities with waveguides, photons emitted from the nanotubes can be guided over the chip and to external components such as optical fibers. The nanophotonic structure also allows us to take advantage of quantum electrodynamical effects, where the acceleration of the emission rate would increase the quantum efficiency.

5.7 Summary

In summary, we have described the unique optical properties of individual SWCNTs. SWCNTs host strongly bound 1D excitons, whose transition energies are inversely proportional to the CNT diameter and can be further tuned in different dielectric environments. Such transitions and their radiative decay can be probed by Raman scattering, light absorption, PL or Rayleigh scattering. Inter-tube coupling in Double-walled carbon nanotubes or hetero-nanotube has been explored. Furthermore, photonic structures have been coupled with a tube for further scaling and increased functionality. However, this is only the beginning of the study and of the application of nanotube photonics. As the production of pure CNTs advances, we expect to see applications in electrically pumped solid-state nanoscale light sources and lasers, CNT light guides and nonlinear devices.

Acknowledgements The authors thank Prof. Shohei Chiashi, Prof. Rong Xiang & Prof. Keigo Otsuka at the University of Tokyo, Prof. Taiki Inoue at Osaka University, and Prof. Yuhei Miyauchi at Kyoto University for discussions and for offering the original data. Part of this work was supported by JSPS KAKENHI Grant Numbers JP18H05329, JP20H00220, JP20H02558 and by JST, CREST Grant Number JPMJCR20B5, Japan as well as MIC (SCOPE 191503001). Kahui Liu acknowledged the supports from the National Natural Science Foundation of China (52025023, 51991340, and 51991342) and Beijing Natural Science Foundation (JQ19004)

References

1. S. Iijima, Nature **354**, 56–58 (1991)
2. S. Iijima, T. Ichihashi, Nature **363**, 603–605 (1993)
3. A.M. Rao, E. Richter, S. Bandow, B. Chase, P.C. Eklund, K.A. Williams, S. Fang, K.R. Subbaswamy, M. Menon, A. Thess, R.E. Smalley, G. Dresselhaus, M.S. Dresselhaus, Science **275**, 187–191 (1997)
4. H. Kataura, Y. Kumazawa, Y. Maniwa, I. Umezu, S. Suzuki, Y. Ohtsuka, Y. Achiba, Synth. Met. **103**, 2555–2558 (1999)
5. A. Jorio, R. Saito, J.H. Hafner, C.M. Lieber, M. Hunter, T. McClure, G. Dresselhaus, M.S. Dresselhaus, Phys. Rev. Lett. **86**, 1118–1121 (2001)
6. S.M. Bachilo, M.B. Strano, C. Kittrell, R.H. Hauge, R.E. Smalley, R.B. Weisman, Science **298**, 2361–2366 (2002)
7. M.S. Arnold, A.A. Green, J.F. Hulvat, S.I. Stupp, M.C. Hersam, Nat. Nanotechnol. **1**, 60–65 (2006)
8. X. Tu, S. Manohar, A. Jagota, M. Zheng, Nature **460**, 250–253 (2009)
9. H. Liu, D. Nishide, T. Tanaka, H. Kataura, Nat. Commun. **2**, 1–8 (2011)
10. J. Lefebvre, Y. Homma, P. Finnie, Phys. Rev. Lett. **90**, 217401 (2003)
11. R. Miura, S. Imamura, R. Ohta, A. Ishii, X. Liu, T. Shimada, S. Iwamoto, Y. Arakawa, Y.K. Kato, Nat. Commun. **5**, 5580 (2014)
12. F. Wang, M.Y. Sfeir, L. Huang, X.M. Huang, Y. Wu, J. Kim, J. Hone, S. O'Brien, L.E. Brus, T.F. Heinz, Phys. Rev. Lett. **96**, 167401 (2006)
13. K. Liu, J. Deslippe, F. Xiao, R.B. Capaz, X. Hong, S. Aloni, A. Zettl, W. Wang, X. Bai, S.G. Louie, E. Wang, F. Wang, Nat. Nanotechnol. **7**, 325–329 (2012)
14. S. Reich, C. Thomsen, J. Maultzsch, *Carbon Nanotubes: Basic Concepts and Physical Properties* (Wiley-VCH, Weinheim, Germany, 2004)

15. R. Saito, G. Dresselhaus, M.S. Dresselhaus, *Physical Properties of Carbon Nanotubes* (Imperial College Press, London, 1998)
16. F. Wang, G. Dukovic, L.E. Brus, T.F. Heinz, Science **308**, 838–841 (2005)
17. J. Lefebvre, P. Finnie, Nano Lett. **8**, 1890–1895 (2008)
18. Y. Toyozawa, *Optical Processes in Solids* (Cambridge Univ, Press, 2003)
19. T. Ando, J. Phys. Soc. Jpn. **66**, 1066–1073 (1997)
20. K.H. Liu, X.P. Hong, S. Choi, C.H. Jin, R.B. Capaz, J. Kim, W.L. Wang, X.D. Bai, S.G. Louie, E.G. Wang, F. Wang, P. Natl, Acad. Sci. **111**, 7564–7569 (2014)
21. J.-C. Blancon, M. Paillet, H.N. Tran, X.T. Than, S.A. Guebrou, A. Ayari, A.S. Miguel, N.-M. Phan, A.-A. Zahab, J.-L. Sauvajol, N.D. Fatti, F. Vallèe, Nat. Commun. **4**, 2542 (2013)
22. M.Y. Sfeir, F. Wang, L.M. Huang, C.C. Chuang, J. Hone, S.P. O'Brien, T.F. Heinz, L.E. Brus, Science **306**, 1540–1543 (2004)
23. M.Y. Sfeir, T. Beetz, F. Wang, L.M. Huang, X.M.H. Huang, M.Y. Huang, J. Hone, S. O'Brien, J.A. Misewich, T.F. Heinz, L.J. Wu, Y.M. Zhu, L.E. Brus, Science **312**, 554–556 (2006)
24. M.S. Dresselhaus, G. Dresselhaus, R. Saito, A. Jorio, Phys. Rep. **409**, 47–99 (2005)
25. M.J. O'Connell, S.M. Bachilo, C.B. Huffman, V.C. Moore, M.S. Strano, E.H. Haroz, K.L. Rialon, P.J. Boul, W.H. Noon, C. Kittrell, J.P. Ma, R.H. Hauge, R.B. Weisman, R.E. Smalley, Science **297**, 593–596 (2002)
26. M. Freitag, Y. Martin, J.A. Misewich, R. Martel, P.H. Avouris, Nano Lett. **3**, 1067–1071 (2003)
27. J. Chen, V. Perebeinos, M. Freitag, J. Tsang, Q. Fu, J. Liu, P. Avouris, Science **310**, 1171–1174 (2005)
28. V.V. Didenko, V.C. Moore, D.S. Baskin, R.E. Smalley, Nano Lett. **5**, 1563–1567 (2005)
29. J. Lefebvre, P. Finnie, Phys. Rev. Lett. **98**, 167406 (2007)
30. D. Mann, Y.K. Kato, A. Kinkhabwala, E. Pop, J. Cao, X. Wang, L. Zhang, Q. Wang, J. Guo, H. Dai, Nat. Nanotechnol. **2**, 33–38 (2007)
31. N.M. Gabor, Z. Zhong, K. Bosnick, J. Park, P.L. McEuen, Science **325**, 1367–1371 (2009)
32. R.B. Weisman, S.M. Bachilo, Nano Lett. **3**, 1235 (2003)
33. J. Lefebvre, J.M. Fraser, Y. Homma, P. Finnie, Appl. Phys. A **78**, 1107 (2004)
34. Y. Ohno, S. Iwasaki, Y. Murakami, S. Kishimoto, S. Maruyama, T. Mizutani, Phys. Rev. B **73**, 235427 (2006)
35. S. Chiashi, S. Watanabe, T. Hanashima, Y. Homma, Nano Lett. **8**, 3097 (2008)
36. C.F. Bohren, D.R. Huffman, *Absorption and Scattering of Light by Small Particles* (John Wiley & Sons, 2008)
37. Z.H. Yu, L. Brus, J. Phys. Chem. B **105**, 1123–1134 (2001)
38. E. Malic, M. Hirtschulz, F. Milde, Y. Wu, J. Maultzsch, T.F. Heinz, A. Knorr, S. Reich, Phys. Rev. B **77**, 045432 (2008)
39. H. Zeng, H. Zhao, F.C. Zhang, X. Cui, Phys. Rev. Lett. **102**, 136406 (2009)
40. D.Y. Joh, L.H. Herman, S.Y. Ju, J. Kinder, M.A. Segal, J.N. Johnson, G.K. Chan, J. Park, Nano Lett. **11**, 1–7 (2011)
41. J. Lefebvre, P. Finnie, Nano Res. **4**, 788–794 (2011)
42. W. Wu, J. Yue, X. Lin, D. Li, F. Zhu, X. Yin, J. Zhu, J. Wang, J. Zhang, Y. Chen, X. Wang, T. Li, Y. He, X. Dai, P. Liu, Y. Wei, J. Wang, W. Zhang, Y. Huang, L. Fan, L. Zhang, Q. Li, S. Fan, K. Jiang, Nano Res. **8**, 2721–2732 (2015)
43. F. Yang, X. Wang, D. Zhang, J. Yang, D. Luo, Z. Xu, J. Wei, J.-Q. Wang, Z. Xu, F. Peng, X. Li, R. Li, Y. Li, M. Li, X. Bai, F. Ding, Y. Li, Nature **510**, 522–524 (2014)
44. S. Zhang, L. Kang, X. Wang, L. Tong, L. Yang, Z. Wang, K. Qi, S. Deng, Q. Li, X. Bai, F. Ding, J. Zhang, Nature **543**, 234–238 (2017)
45. Y.X. Bai, R.F. Zhang, X. Ye, Z.X. Zhu, H.H. Xie, B.Y. Shen, D.L. Cai, B.F. Liu, C.X. Zhang, Z. Jia, S.L. Zhang, X.D. Li, F. Wei, Nat. Nanotechnol. **13**, 589–595 (2018)
46. D. Christofilos, J.C. Blancon, J. Arvanitidis, A.S. Miguel, A. Ayari, N. Del Fatti, F. Vallée, J. Phys. Chem. C **3**, 1176–1181 (2012)
47. Y. Wu, M. Huang, F. Wang, X.H. Huang, S. Rosenblatt, L. Huang, H. Yan, S.P. O'Brien, J. Hone, T.F. Heinz, Nano Lett. **8**, 4158–4161 (2008)

48. M.F. Islam, D.E. Milkie, C.L. Kane, A.G. Yodh, J.M. Kikkawa, Phys. Rev. Lett. **93**, 037404 (2004)
49. Y. Murakami, E. Einarsson, T. Edamura, S. Maruyama, Phys. Rev. Lett. **94**, 087402 (2005)
50. Y. Miyauchi, M. Oba, S. Maruyama, Phys. Rev. B **74**, 205440 (2006)
51. F. Vialla, C. Roquelet, B. Langlois, G. Delport, S.M. Santos, E. Deleporte, P. Roussignol, C. Delalande, C. Voisin, J.-S. Lauret, Phys. Rev. Lett. **111**, 137402 (2013)
52. J.A. Fagan, C.Y. Khripin, C.A. Silvera Batista, J.R. Simpson, E.H. Hroz, A.R. Hight Walker, M. Zheng, Adv. Mater. **26**, 2800–2804 (2014)
53. Y. Yomogida, T. Tanaka, M. Zhang, M. Yudasaka, X. Wei, H. Kataura, Nat. Commun. **7**, 1–8 (2016)
54. G. Ao, C.Y. Khripin, M. Zheng, J. Am. Chem. Soc. **136**, 103183–10392 (2014)
55. Y. Ohno, S. Kishimoto, T. Mizutani, T. Okazaki, H. Shinohara, Appl. Phys. Lett. **84**, 1368–1370 (2004)
56. E.J. Lee, K. Balasubramanian, J. Dorfmller, R. Vogelgesang, N. Fu, A. Mews, M. Burghard, K. Kern, Small **3**, 2038–2042 (2007)
57. M. Barkelid, G.A. Steele, V. Zwiller, Nano Lett. **12**, 5649–5653 (2012)
58. M. Barkelid, V. Zwiller, Nat. Photon. **8**, 47–51 (2013)
59. S. Berciaud, L. Cognet, P. Poulin, R.B. Weisman, B. Lounis, Nano Lett. **7**, 1203–1207 (2007)
60. F. Wang, D.J. Cho, B. Kessler, J. Deslippe, P.J. Schuck, S.G. Louie, A. Zettl, T.F. Heinz, Y.R. Shen, Phys. Rev. Lett. **99**, 227401 (2007)
61. K. Liu, X. Hong, Q. Zhou, C. Jin, J. Li, W. Zhou, J. Liu, E. Wang, A. Zettl, F. Wang, Nat. Nanotechnol. **8**, 917–922 (2013)
62. L.H. Herman, C.J. Kim, Z.H. Wang, M.H. Jo, J. Park, Appl. Phys. Lett. **101**, 123102 (2012)
63. F. Yao, C. Liu, C. Chen, S. Zhang, Q. Zhao, F. Xiao, M. Wu, J. Li, P. Gao, J. Zhao, X. Bai, S. Maruyama, D. Yu, E. Wang, Z. Sun, J. Zhang, F. Wang, K. Liu, Nat. Commun. **9**, 3387 (2018)
64. H.B. Chu, R.L. Cui, J.Y. Wang, J.A. Yang, Y. Li, Carbon **49**, 1182–1188 (2011)
65. R. Zhang, Y. Zhang, Q. Zhang, H. Xie, H. Wang, J. Nie, Q. Wen, F. Wei, Nat. Commun. **4**, 1727–1735 (2013)
66. J. Wang, T. Li, B. Xia, X. Jin, H. Wei, W. Wu, Y. Wei, J. Wang, P. Liu, L. Zhang, Nano Lett. **14**, 3527–3533 (2014)
67. S. Chaudhary, J.H. Kim, K.V. Singh, M. Ozkan, Nano Lett. **4**, 2415–2419 (2004)
68. K. Balasubramanian, Y. Fan, M. Burghard, K. Kern, M. Friedrich, U. Wannek, A. Mews, Appl. Phys. Lett. **84**, 2400–2402 (2004)
69. A.W. Tsen, L.A. Donev, H. Kurt, L.H. Herman, J. Park, Nat. Nanotechnol. **4**, 108–113 (2009)
70. F. Wang, S. Wang, F. Yao, H. Xu, N. Wei, K. Liu, L.-M. Peng, ACS Nano **10**, 9595–9601 (2016)
71. Q. Zhao, F. Yao, Z. Wang, S. Deng, L. Tong, K. Liu, J. Zhang, Adv. Mater. **29**, 1701959 (2017)
72. A. Ishii, M. Yoshida, Y.K. Kato, Phys. Rev. B **91**, 125427 (2015)
73. S. Maruyama, R. Kojima, Y. Miyauchi, S. Chiashi, M. Kohno, Chem. Phys. Lett. **360**, 229 (2002)
74. L. Huang, X. Cui, B. White, S.P. O'Brien, J. Phys. Chem. B **108**, 16451 (2004)
75. P.H. Tan, A.G. Rozhin, T. Hasan, P. Hu, V. Scardaci, W.I. Milne, A.C. Ferrari, Phys. Rev. Lett. **99**, 137402 (2007)
76. L. Yang, M.P. Anantram, J. Han, J.P. Lu, Phys. Rev. B **60**, 13874 (1999)
77. M. Huang, Y. Wu, B. Chandra, H. Yan, Y. Shan, T.F. Heinz, J. Hone, Phys. Rev. Lett. **100**, 136803 (2008)
78. A. G. Souza Filho, N. Kobayashi, J. Jiang, A. Gr'uneis, R. Saito, S. B. Cronin, J. Mendes Filho, G. G. Samsonidze, G. Dresselhaus, M. S. Dresselhaus, Phys. Rev. Lett. **95**, 217403 (2005)
79. P. Koskinen, Phys. Rev. B **82**, 193409 (2010)
80. S. Bandow, S. Asaka, Y. Saito, A.M. Rao, L. Grigorian, E. Richter, P.C. Eklund, Phys. Rev. Lett. **80**, 3779 (1998)

81. S.M. Bachilo, L. Balzano, J.E. Herrera, F. Pompeo, D.E. Resasco, R.B. Weisman, J. Am. Chem. Soc. **125**, 11186 (2003)
82. Y. Miyauchi, S. Chiashi, Y. Murakami, Y. Hayashida, S. Maruyama, Chem. Phys. Lett. **387**, 198 (2004)
83. R. Rao, D. Liptak, T. Cherukuri, B.I. Yakobson, B. Maruyama, Nat. Mater. **11**, 213 (2012)
84. A. Jorio, C. Fantini, M.A. Pimenta, R.B. Capaz, G.G. Samsonidze, G. Dresselhaus, M.S. Dresselhaus, J. Jiang, N. Kobayashi, A. Grüneis, R. Saito, Phys. Rev. B **71**, 075401 (2005)
85. S. Chiashi, K. Kono, D. Matsumoto, J. Shitaba, N. Homma, A. Beniya, T. Yamamoto, Y. Homma, Phys. Rev. B **91**, 155415 (2015)
86. K. Otsuka, S. Yamamoto, T. Inoue, B. Koyano, H. Ukai, R. Yoshikawa, R. Xiang, S. Chiashi, S. Maruyama, ACS Nano **12**, 3994–4001 (2018)
87. B. Hou, C. Wu, T. Inoue, S. Chiashi, R. Xiang, S. Maruyama, Carbon **119**, 502–510 (2017)
88. F. Tuinstra, J.L. Koenig, J. Chem. Phys. **53**, 1126–1130 (1970)
89. A. Jorio, M.A. Pimenta, A.G. Souza, G.G. Samsonidze, A.K. Swan, M.S. Unlu, B.B. Goldberg, R. Saito, G. Dresselhaus, M.S. Dresselhaus, Phys. Rev. Lett. **90**, 107403 (2003)
90. S. Lebedkin, F. Hennrich, T. Skipa, M.M. Kappes, J. Phys. Chem. B **107**, 1949–1956 (2003)
91. Y. Kobayashi, T. Yamashita, Y. Ueno, O. Niwa, Y. Homma, T. Ogino, Chem. Phys. Lett. **386**, 153–157 (2004)
92. Y. Piao, J.R. Simpson, J.K. Streit, G. Ao, M. Zheng, J.A. Fagan, A.R.H. Walker, ACS Nano **10**, 5252–5259 (2016)
93. H.B. Son, Y. Hori, S.G. Chou, D. Nezich, G.G. Samsonidze, G. Dresselhaus, M.S. Dresselhaus, E.B. Barros, Appl. Phys. Lett. **85**, 4744–4746 (2004)
94. V. Avramenko, S.B. Rochal, Phys. Rev. B **87**, 035407 (2013)
95. X. Blase, L.X. Benedict, E.L. Shirley, S.G. Louie, Phys. Rev. Lett. **72**, 1878–1881 (1994)
96. A. Jorio, A.G. Souza, G. Dresselhaus, M.S. Dresselhaus, A.K. Swan, M.S. Unlu, B.B. Goldberg, M.A. Pimenta, J.H. Hafner, C.M. Lieber, R. Saito, Phys. Rev. B **65**, 155412 (2002)
97. H. Telg, J.G. Duque, M. Staiger, X. Tu, F. Hennrich, M.M. Kappes, M. Zheng, J. Maultzsch, C. Thomsen, S.K. Doorn, ACS Nano **6**, 904–911 (2012)
98. H. Farhat, H. Son, G. Samsonidze, S. Reich, M. Dresselhaus, J. Kong, Phys. Rev. Lett. **99**, 145506 (2007)
99. S.D.M. Brown, A. Jorio, P. Corio, M.S. Dresselhaus, G. Dresselhaus, R. Saito, K. Kneipp, Phys. Rev. B **63**, 155414 (2001)
100. J.S. Park, K. Sasaki, R. Saito, W. Izumida, M. Kalbac, H. Farhat, G. Dresselhaus, M.S. Dresselhaus, Phys. Rev. B **80**, 081402(R) (2009)
101. K. Nguyen, A. Gaur, M. Shim, Phys. Rev. Lett. **98**, 145504 (2007)
102. B. Hatting, S. Heeg, K. Ataka, J. Heberle, F. Hennrich, M.M. Kappes, R. Krupke, S. Reich, Phys. Rev. B **87**, 165442 (2013)
103. M.A. Pimenta, A. Marucci, S.A. Empedocles, M.G. Bawendi, E.B. Hanlon, A.M. Rao, P.C. Eklund, R.E. Smalley, G. Dresselhaus, M.S. Dresselhaus, Phys. Rev. B **58**, R16016–R16019 (1998)
104. J.G. Duque, H. Chen, A.K. Swan, A.P. Shreve, S. Kilina, S. Tretiak, X. Tu, M. Zheng, S.K. Doorn, ACS Nano **5**, 5233–5241 (2011)
105. G.D. Mahan, Phys. Rev. B **65**, 2354027 (2002)
106. V. Perebeinos, J. Tersoff, Phys. Rev. B **79**, 241409 (2009)
107. S. Chiashi, Y. Homma, S. Maruyama, Handbook of Carbon Nanomaterials **10**, 49–73 (2019)
108. J. Jiang, R. Saito, K. Sato, J.S. Park, G.G. Samsonidze, A. Jorio, G. Dresselhaus, M.S. Dresselhaus, Phys. Rev. B **75**, 035405 (2007)
109. C. Fantini, A. Jorio, M. Souza, M. S. Strano, M. S. Dresselhaus, and M. A. Pimenta, Phys. Rev. Lett. **93**, 147406, (2004)
110. P. T. Araujo, S. K. Doorn, S. Kilina, S. Tretiak, E. Einarsson, S. Maruyama, H. Chacham, M. A. Pimenta, A. Jorio, Phys. Rev. Lett. **98**, 067401, (2007)
111. H. Telg, J. Maultzsch, S. Reich, F. Hennrich, C. Thomsen, phys. stat. sol. (b) **243**, 3181–3185, (2006)

112. T. Michel, M. Paillet, J. Meyer, V. Popov, L. Henrard, P. Poncharal, A. Zahab, J. Sauvajol, phys. stat. sol. (b) **244**, 3986–3991 (2007)
113. P. T. Araujo, A. Jorio, phys. stat. sol. (b) **245**, 2201–2204 (2008)
114. J.C. Charlier, P. Lambin, Phys. Rev. B **57**, R15037–R15039 (1998)
115. G.G. Samsonidze, R. Saito, N. Kobayashi, A. Gruneis, J. Jiang, A. Jorio, S.G. Chou, G. Dresselhaus, M.S. Dresselhaus, Appl. Phys. Lett. **85**, 5703–5705 (2004)
116. L. Yang, J. Han, Phys. Rev. Lett. **85**, 154–157 (2000)
117. K.H. Liu, C.H. Jin, X.P. Hong, J. Kim, A. Zettl, E.G. Wang, F. Wang, Nat. Phys. **10**, 737–742 (2014)
118. K. Liu, X. Hong, M. Wu, F. Xiao, W. Wang, X. Bai, J.W. Ager, S. Aloni, A. Zettl, E. Wang, Nat. Commun. **4**, 1–6 (2013)
119. D.I. Levshov, H.N. Tran, T. Michel, T.T. Cao, V.C. Nguyen, R. Arenal, V.N. Popov, J.L. Sauvajol, A.A. Zahab, M. Paillet, phys. stat. sol. (b) **254**, 1700251 (2017)
120. H.N. Tran, J.-C. Blancon, R. Arenal, R. Parret, A.A. Zahab, A. Ayari, F. Vallee, N. Del Fatti, J.-L. Sauvajol, M. Paillet, Phys. Rev. B **95**, 205411 (2017)
121. R. Bonnet, A. Lherbier, C. Barraud, M.L. Della Rocca, P. Lafarge, J.-C. Charlier, Sci. Rep. **6**, 19701 (2016)
122. S. Zhao, P. Moon, Y. Miyauchi, T. Nishihara, K. Matsuda, M. Koshino, R. Kitaura, Phys. Rev. Lett. **124**, 106101 (2020)
123. D.V. Chalin, S.B. Rochal, Phys. Rev. B **102**, 115426 (2020)
124. R. Moradian, S. Azadi, H. Refii-Tabar, J. Phys. Condens. Mat. **19**, 176209 (2007)
125. D.I. Levshov, R. Parret, H.-N. Tran, T. Michel, T.T. Cao, R. Arenal, V.N. Popov, S.B. Rochal, J.-L. Sauvajol, A.-A. Zahab, Phys. Rev. B **96**, 195410 (2017)
126. R. Xiang, T. Inoue, Y. Zheng, A. Kumamoto, Y. Qian, Y. Sato, M. Liu, D. Tang, D. Gokhale, J. Guo, K. Hisama, S. Yotsumoto, T. Ogamoto, H. Arai, Y. Kobayashi, H. Zhang, B. Hou, A. Anissimov, M. Maruyama, Y. Miyata, S. Okada, S. Chiashi, Y. Li, E.I. Kauppinen, Y. Ikuhara, K. Suenaga, S. Maruyama, Science **367**, 537–542 (2020)
127. M. Bockrath, D.H. Cobden, J. Lu, A.G. Rinzler, R.E. Smalley, L. Balents, P.L. McEuen, Nature **397**, 598–601 (1999)
128. V.V. Deshpande, M. Bockrath, L.I. Glazman, A. Yacoby, Nature **464**, 209–216 (2010)
129. H. Ishii, H. Kataura, H. Shiozawa, H. Yoshioka, H. Otsubo, Y. Takayama, T. Miyahara, S. Suzuki, Y. Achiba, M. Nakatake, T. Narimura, M. Higashiguchi, K. Shimada, H. Namatame, M. Taniguchi, Nature **426**, 540–544 (2003)
130. Z. Shi, X. Hong, H.A. Bechtel, B. Zeng, M.C. Martin, K. Watanabe, T. Taniguchi, Y.-R. Shen, F. Wang, Nat. Photonics. **9**, 515–519 (2015)
131. S. Wang, S. Zhao, Z. Shi, F. Wu, Z. Zhao, L. Jiang, K. Watanabe, T. Taniguchi, A. Zettl, C. Zhou, F. Wang, Nat. Mater. **1**, 6 (2020)
132. J. Kong, H.T. Soh, A.M. Cassell, C.F. Quate, H. Dai, Nature **395**, 878 (1998)
133. J.A. Misewich, R. Martel, P. Avouris, J.C. Tsang, S. Heinze, J. Tersoff, Science **300**, 783 (2003)
134. F. Xia, M. Steiner, Y.-M. Lin, P. Avouris, Nat. Nanotech. **3**, 609 (2008)
135. T. Mueller, M. Kinoshita, M. Steiner, V. Perebeinos, A.A. Bol, D.B. Farmer, P. Avouris, Nat. Nanotech. **5**, 27 (2010)
136. A. Graf, L. Tropf, Y. Zakharko, J. Zaumseil, M.C. Gather, Nat. Commun. **7**, 13078 (2016)
137. R. Watahiki, T. Shimada, P. Zhao, S. Chiashi, S. Iwamoto, Y. Arakawa, S. Maruyama, Y.K. Kato, Appl. Phys. Lett. **101**, 141124 (2012)
138. S. Imamura, R. Watahiki, R. Miura, T. Shimada, Y.K. Kato, Appl. Phys. Lett. **102**, 161102 (2013)
139. A. Noury, X.L. Roux, L. Vivien, N. Izard, Nanotechnology **26**, 345201 (2015)
140. F. Pyatkov, V. Fütterling, S. Khasminskaya, B.S. Flavel, F. Hennrich, M.M. Kappes, R. Krupke, W.H.P. Pernice, Nat. Photon. **10**, 420 (2016)
141. S. Moritsubo, T. Murai, T. Shimada, Y. Murakami, S. Chiashi, S. Maruyama, Y.K. Kato, Phys. Rev. Lett. **104**, 247402 (2010)

142. I. Sarpkaya, Z. Zhang, W. Walden-Newman, X. Wang, J. Hone, C.W. Wong, S. Strauf, Nat. Commun. **4**, 2152 (2013)
143. M. Barkelid, V. Zwiller, Nat. Photon. **8**, 47 (2014)
144. M. Eichenfield, R. Camacho, J. Chan, K.J. Vahala, O. Painter, Nature **459**, 550 (2009)
145. Y. Gong, B. Ellis, G. Shambat, T. Sarmiento, J.S. Harris, J. Vuckovi'c, Opt. Express **18**, 8781 (2010)
146. R. Ohta, Y. Ota, M. Nomura, N. Kumagai, S. Ishida, S. Iwamoto, Y. Arakawa, Appl. Phys. Lett. **98**, 173104 (2011)
147. J. Riedrich-Möller, L. Kipfstuhl, C. Hepp, E. Neu, C. Pauly, F. Mücklich, A. Baur, M. Wandt, S. Wolff, M. Fischer, S. Gsell, M. Schreck, C. Becher, Nat. Nanotech. **7**, 69 (2012)
148. B.J.M. Hausmann, B.J. Shields, Q. Quan, Y. Chu, N.P. de Leon, R. Evans, M.J. Burek, A.S. Zibrov, M. Markham, D.J. Twitchen, H. Park, M.D. Lukin, M. Loncar, Nano Lett. **13**, 5791 (2013)

Index

T. Yatsui (ed.), *Progress in Nanophotonics 6*, Nano-Optics and Nanophotonics,
https://doi.org/10.1007/978-3-030-71516-8